Evaluation of the Voluntary National Tests, Year 2

FINAL REPORT

Committee on the Evaluation of the Voluntary National Tests, Year 2

Lauress L. Wise, Richard J. Noeth, and Judith A. Koenig, *Editors*

Board on Testing and Assessment

Commission on Behavioral and Social Sciences and Education

National Research Council

NATIONAL ACADEMY PRESS
Washington, DC

NATIONAL ACADEMY PRESS • 2101 Constitution Avenue, N.W. • Washington, D.C. 20418

NOTICE: The project that is the subject of this report was approved by the Governing Board of the National Research Council, whose members are drawn from the councils of the National Academy of Sciences, the National Academy of Engineering, and the Institute of Medicine. The members of the committee responsible for the report were chosen for their special competences and with regard for appropriate balance.

The study was supported by Contract/Grant No. RJ97184001 between the National Academy of Sciences and the U.S. Department of Education. Any opinions, findings, conclusions, or recommendations expressed in this publication are those of the author(s) and do not necessarily reflect the view of the organizations or agencies that provided support for this project.

International Standard Book Number 0-309-06788-X

Additional copies of this report are available from:
National Academy Press
2101 Constitution Avenue NW
Washington, DC 20418
Call 800-624-6242 or 202-334-3313 (in the Washington Metropolitan Area).

This report is also available on line at http://www.nap.edu

Printed in the United States of America

Suggested citation: National Research Council (1999) *Evaluation of the Voluntary National Tests, Year 2: Final Report.* Committee on The Evaluation of the Voluntary National Tests, Year 2. Lauress L. Wise, Richard J. Noeth, and Judith A. Keonig, editors. Board on Testing and Assessment, Commission on Behavioral and Social Sciences and Education. Washington, DC: National Academy Press.

THE NATIONAL ACADEMIES

National Academy of Sciences
National Academy of Engineering
Institute of Medicine
National Research Council

The **National Academy of Sciences** is a private, nonprofit, self-perpetuating society of distinguished scholars engaged in scientific and engineering research, dedicated to the furtherance of science and technology and to their use for the general welfare. Upon the authority of the charter granted to it by the Congress in 1863, the Academy has a mandate that requires it to advise the federal government on scientific and technical matters. Dr. Bruce M. Alberts is president of the National Academy of Sciences.

The **National Academy of Engineering** was established in 1964, under the charter of the National Academy of Sciences, as a parallel organization of outstanding engineers. It is autonomous in its administration and in the selection of its members, sharing with the National Academy of Sciences the responsibility for advising the federal government. The National Academy of Engineering also sponsors engineering programs aimed at meeting national needs, encourages education and research, and recognizes the superior achievements of engineers. Dr. William A. Wulf is president of the National Academy of Engineering.

The **Institute of Medicine** was established in 1970 by the National Academy of Sciences to secure the services of eminent members of appropriate professions in the examination of policy matters pertaining to the health of the public. The Institute acts under the responsibility given to the National Academy of Sciences by its congressional charter to be an adviser to the federal government and, upon its own initiative, to identify issues of medical care, research, and education. Dr. Kenneth I. Shine is president of the Institute of Medicine.

The **National Research Council** was organized by the National Academy of Sciences in 1916 to associate the broad community of science and technology with the Academy's purposes of furthering knowledge and advising the federal government. Functioning in accordance with general policies determined by the Academy, the Council has become the principal operating agency of both the National Academy of Sciences and the National Academy of Engineering in providing services to the government, the public, and the scientific and engineering communities. The Council is administered jointly by both Academies and the Institute of Medicine. Dr. Bruce M. Alberts and Dr. William A. Wulf are chairman and vice chairman, respectively, of the National Research Council.

COMMITTEE ON THE EVALUATION OF THE VOLUNTARY NATIONAL TESTS, YEAR 2

LAURESS L. WISE (*Chair*), Human Resources Research Organization, Alexandria, Virginia
ADRIENNE BAILEY, Consultant, Chicago, Illinois
THOMAS COONEY, Department of Mathematics Education, University of Georgia
JOHN T. GUTHRIE, Department of Human Development, University of Maryland
ANNE HAFNER, Charter School of Education, California State University, Los Angeles
ROBERT M. HAUSER, Department of Sociology, University of Wisconsin, Madison
VONDA L. KIPLINGER, Colorado Department of Education, Denver
MARJORIE Y. LIPSON, Department of Education, University of Vermont
ALFRED MANASTER, Department of Mathematics, University of California, San Diego
NANCY S. PETERSEN, ACT, Inc., Iowa City, Iowa

RICHARD J. NOETH, *Study Director*
JUDITH A. KOENIG, *Program Officer*
KIMBERLY D. SALDIN, *Senior Project Assistant*

Preface

President Clinton's 1997 proposal to create voluntary national tests in fourth grade reading and eighth grade mathematics did much to heighten the ongoing national debate about testing in America's schools. The National Research Council has been asked by Congress and the White House to play a key role in this debate by conducting several interrelated studies to provide advice on these important assessment issues.

Through its Board on Testing and Assessment (BOTA), the NRC issued three significant studies in 1998 to provide such advice: *Evaluation of the Voluntary National Tests: Phase 1 Report*, *Uncommon Measures: Equivalence and Linkage Among Educational Tests*, and *High Stakes: Testing for Tracking, Promotion, and Graduation.* BOTA has continued this important work this year by conducting two further studies. These include the present year 2 Voluntary National Tests (VNT) evaluation described in this report and in *Embedding Questions: The Pursuit of a Common Measure in Uncommon Tests*.

The NRC's approach to the second year of its VNT evaluation differed from the first year in several ways. First, the work was conducted by means of a traditional National Research Council committee rather than by co-principal investigators. The committee of ten experts in reading, mathematics, assessment, educational policy, and test use allowed the NRC to bring a wider range of expertise to the planning and conduct of the evaluation and reduced reliance on outside experts. The use of a study committee was related to the second change in the VNT evaluation, which is an expanded scope that included a continued principal focus on the quality of the items being developed, technical issues in test development, and inclusion and accommodation issues, as well as the tests' purpose and how the VNT would be used.

This project would not have been possible without the generosity of many individuals and the contributions of several institutions.

Staff from the National Assessment Governing Board (NAGB), under the leadership of Roy Truby, executive director, and the NAGB prime contractor, the American Institutes for Research (AIR), with Steve Ferrara and Archie LaPointe's guidance, were a valuable source of information and data on the design and development of the Voluntary National Tests. Sharif Shakrani, Steve Gorman, Raymond Fields, Mary Lyn Bourque, and Mary Crovo of NAGB and Steve Klein, Clayton Best, Ruth Childs, and Terry Salinger of AIR provided us with important information on numerous occasions. We benefited tremendously by attending and learning from discussions at meetings of the National Assessment Governing Board and meetings of its contractors; we thank them for opening their meetings to us and for sharing their knowledge and perspectives. We extend thanks to the staff of the cognitive laboratories and of Harcourt Brace Educational Measurement and Riverside Publishing for access to important information and their perspectives throughout the course of our work.

We relied heavily on the input and advice of a cadre of testing and disciplinary experts, who provided helpful and insightful presentations at our workshops: Jamal Abedi, University of California; Pamela Beck, University of California; Jeffrey Choppin, Benjamin Banneker Academic High School; Gregory Cizek, University of Toledo; Jonathan Dings, Boulder Valley School District; Gretchen Glick, Defense Manpower Data Center; Anna Graeber, University of Maryland; Lorraine McDonnell, University of California; Rosemarie Montgomery, Hatboro, Pennsylvania; Lorrie Shepard, University of Colorado; Gale Sinatra, University of Utah; John Tanner, Delaware Department of Education; Wendy Yen, CTB/McGraw-Hill, and Catherine Yohe, Williamsburg Middle School. Our work was enriched by the stimulating intellectual exchange at the meeting and item quality workshop to which they contributed greatly.

Carolyn Harris, Gene Hoffman, Sunny Sipes, Don Smith, and Art Thacker of Human Resources Research Organization provided important help and perspective throughout. They attended and reported on workshops, cognitive laboratories, bias review sessions, public hearings, and were valuable members of the evaluation team.

This report has been reviewed in draft form by individuals chosen for their diverse perspectives and technical expertise, in accordance with procedures approved by the Report Review Committee of the National Research Council (NRC). The purpose of this independent review is to provide candid and critical comments that will assist the institution in making the published report as sound as possible and to ensure that the report meets institutional standards for objectivity, evidence, and responsiveness to the study charge. The review comments and draft manuscript remain confidential to protect the integrity of the deliberative process.

We thank the following individuals for their participation in the review of this report: Daniel Heilborn, School of Medicine, University of California, San Francisco; Paul Holland, Graduate School of Education, University of California, Berkeley; Lyle V. Jones, L.L. Thurstone Psychometric Laboratory, University of North Carolina, Chapel Hill; Michael J. Kolen, Iowa Testing Programs, University of Iowa; Duncan MacQuarrie, Department of Curriculum and Assessment, Office of the Superintendent of Public Instruction, Washington State Department of Education; Stephen Raudenbush, School of Education, University of Michigan; and Henry W. Riecken, Professor of Behavioral Sciences (emeritus), University of Pennsylvania School of Medicine.

Although the individuals listed above have provided constructive comments and suggestions, it must be emphasized that responsibility for the final content of this report rests entirely with the authoring committee and the institution.

We are grateful to the many individuals at the National Research Council who provided guidance and assistance at many stages of the evaluation and during the preparation of the report. Barbara Torrey, executive director of the Commission on Behavioral and Social Sciences and Education (CBASSE), helped and encouraged our work throughout. We are especially grateful to Eugenia Grohman, associate director for reports of CBASSE, for her advice on structuring the content of the report, for her expert editing of the manuscript, for her wise advice on the exposition of the report's main messages, and for her patient and deft guidance of the report through the NRC report process.

We also are immensely grateful to Stephen Baldwin, Lee Jones, Patricia Morison, and Meryl Bertenthal, staff of the Board on Testing and Assessment (BOTA), who made valuable contributions to our research and report. We express our gratitude to NRC administrative staff Viola Horek and Lisa Alston. We are especially grateful to Kimberly Saldin, who capably and admirably managed the operational aspects of the evaluation—arranging meeting and workshop logistics, producing multiple iterations of drafts and report text, and being available to assist with our requests, however large or small.

We recognize the special contributions of Michael Feuer, BOTA director, and Karen Mitchell, senior program officer. Michael guided the project and, most important, made frequent contributions to the discussion and the framing of our questions and conclusions. Karen was a principal source of expertise in both the substance and process of the evaluation, and she provided continuous liaison between us and the staff of NAGB and AIR.

Lauress Wise, Richard Noeth, and Judith Koenig,
Editors
Committee on the Evaluation of the
Voluntary National Tests, Year 2

Contents

Executive Summary

In his 1997 State of the Union address, President Clinton announced a federal initiative to develop tests of 4th-grade reading and 8th-grade mathematics that could be administered on a voluntary basis by states and school districts beginning in spring 1999. The principal purpose of the Voluntary National Tests (VNT) is to provide parents and teachers with systematic and reliable information about the verbal and quantitative skills that students have achieved at two key points in their educational careers. The U.S. Department of Education anticipated that this information would serve as a catalyst for continued school improvement, by focusing parental and community attention on achievement and by providing an additional tool to hold school systems accountable for their students' performance in relation to nationwide standards.

Shortly after initial development work on the VNT, Congress transferred responsibility for VNT policies, direction, and guidelines from the department to the National Assessment Governing Board (NAGB, the governing body for the National Assessment of Educational Progress). Test development activities were to continue, but Congress prohibited pilot and field testing and operational use of the VNT pending further consideration. At the same time, Congress called on the National Research Council (NRC) to assess the VNT development activities. Specifically, the NRC was charged to evaluate:

1. the technical quality of any test items for 4th-grade reading and 8th-grade mathematics;
2. the validity, reliability, and adequacy of developed test items;
3. the validity of any developed design which links test results to student performance levels;
4. the degree to which any developed test items provide valid and useful information to the public;
5. whether the test items are free from racial, cultural, or gender bias;
6. whether the test items address the needs of disadvantaged, limited English proficient, and disabled students; and
7. whether the test items can be used for tracking, graduation, or promotion of students.

Since the evaluation began, the NRC has issued three reports on VNT development: an interim and final report on the first year's work and an interim report earlier on this second year's work. This final report includes the findings and recommendations from the interim report, modified by new information and analysis, and presents our overall conclusions and recommendations regarding the VNT.

Congress must answer the overarching policy question about the VNT: whether development should continue or be terminated. The committee does not take a position either for or against continued development. However, the committee has reached two general conclusions about the development of the VNT. Our first conclusion deals with our overall assessment.

CONCLUSION VNT development is generally on course. A large number of items have been written, and the quality of the items that came through the contractor's development and review process is comparable to the quality of items from NAEP. Plans for pilot testing these items are generally sound.

The committee does have a number of specific recommendations to further improve item and test quality, to improve future cycles of the test development process, to enhance the meaningful inclusion of all students, to provide effective reports of results, and to prevent misuse of test results. However, there is no evidence that the current process should be halted on technical grounds.

Our second general conclusion addresses the potential value of a pilot test of VNT items.

CONCLUSION The planned pilot test of VNT test items presents opportunities for research on a number of important test development topics that will be useful to NAEP and state and local assessment programs even if the VNT is eventually terminated. These research opportunities include: (1) assessing the quality and effectiveness of the VNT's item development, review, and revisions processes; (2) collecting empirical data on the effect of different threats to "linkability;" and (3) assessing the feasibility, effects, and validity of alternative testing accommodations for students with disabilities or limited English proficiency. In addition, the items themselves are likely to be useful for other testing programs.

Finally, the committee offers an overall recommendation to Congress in considering its decision about the VNT.

RECOMMENDATION TO CONGRESS The decision to continue or terminate the VNT should be based on a carefully articulated statement of the expected value and costs of the program, including a detailed examination of underlying assumptions and a delineation of possible unintended outcomes. To the maximum extent possible, research on results from other educational reform efforts should be laid out to support or contradict assumptions in this value-and-cost statement. Information on the likelihood of use by states, districts, and individuals should also be considered in making a decision about the VNT.

1

Introduction and History

In his 1997 State of the Union address, President Clinton announced a federal initiative to develop tests of 4th-grade reading and 8th-grade mathematics that could be administered on a voluntary basis by states and school districts beginning in spring 1999. The call for Voluntary National Tests (VNT) echoed a similar proposal for "America's Test," which the Bush administration offered in 1990. The principal purpose of the VNT, as articulated by the Secretary of the U.S. Department of Education (see, e.g., Riley, 1997), is to provide parents and teachers with systematic and reliable information about the verbal and quantitative skills that students have achieved at two key points in their school lives. The Department of Education anticipates that this information will serve as a catalyst for continued school improvement, by focusing parental and community attention on achievement and by providing an additional tool to hold school systems accountable for their students' performance in relation to nationwide standards.

The proposed VNT has evolved in many ways since January 1997, but the major features were clear in the initial plan. Achievement tests in English reading at the 4th-grade level and in mathematics at the 8th-grade level would be offered to states, school districts, and localities for administration in the spring of each school year. Several other features of the tests were specified:

- The tests would be voluntary: the federal government would prepare but not require them, nor would data on any individual, school, or group be reported to the federal government.
- The tests, each administered in two, 45-minute sessions in a single day, would not be long or detailed enough to provide diagnostic information about individual learning problems. Rather, they would provide reliable information so all students—and their parents and teachers—would know where they are in relation to high national standards. In mathematics results would be linked to scores from the Third International Mathematics and Science Study (TIMSS) to provide comparisons with student performance in other countries.
- The tests would be designed to facilitate linkage with the National Assessment of Educational

Progress (NAEP) and the reporting of individual test performance in terms of the NAEP achievement levels: basic, proficient, and advanced.

• In order to provide maximum preparation and feedback to students, parents, and teachers, sample tests would be circulated in advance, copies of the original tests would be returned with the students' original and correct answers, and all test items would be published on the Internet just after the administration of each test.

Initial plans for the VNT were laid out by the Department of Education and, in late summer 1997, a contract for test development was awarded to a consortium led by the American Institutes for Research (AIR). The original schedule called for development of test specifications for 4th-grade reading and 8th-grade mathematics tests by fall 1997; pilot testing of test items later that year; field testing of test forms early in 1998; and the first test administration in spring 1999. The department also awarded a contract to the National Research Council (NRC) to conduct an evaluation of VNT test development activities.

Subsequent negotiations between the administration and Congress, which culminated in passage of the fiscal 1998 appropriations bill (P.L. 105-78), led to a suspension of test item development (a stop-work order) late in September 1997 and transferred to the National Assessment Governing Board (NAGB, the governing body for NAEP) exclusive authority to oversee the policies, direction, and guidelines for developing the VNT. The law gave NAGB 90 days in which to review the development plan and revise or renegotiate the test development contract.

Congress further instructed NAGB to make four determinations about the VNT:

(1) the extent to which test items selected for use on the tests are free from racial, cultural, or gender bias;
(2) whether the test development process and test items adequately assess student reading and mathematics comprehension in the form most likely to yield accurate information regarding student achievement in reading and mathematics;
(3) whether the test development process and test items take into account the needs of disadvantaged, limited English proficient, and disabled students; and
(4) whether the test development process takes into account how parents, guardians, and students will be appropriately informed about testing content, purpose, and uses.

NAGB negotiated a revised schedule and work plan with AIR. It called for test development over a 3-year period—with pilot testing in March 1999, field testing in March 2000, and operational test administration in March 2001. In addition, the work plan specified a major decision point in fall 1998, which depended on congressional action, and it permitted limited test development activities to proceed through the remainder of the fiscal year, to September 30, 1998.

When the Congress assigned NAGB responsibility for the VNT, it also called on the NRC to evaluate the technical adequacy of test materials. Specifically, it asked the NRC to evaluate:

(1) the technical quality of any test items for 4th-grade reading and 8th-grade mathematics;
(2) the validity, reliability, and adequacy of developed test items;
(3) the validity of any developed design which links test results to student performance levels;
(4) the degree to which any developed test items provide valid and useful information to the public;
(5) whether the test items are free from racial, cultural, or gender bias;

(6) whether the test items address the needs of disadvantaged, limited English proficient, and disabled students; and
(7) whether the test items can be used for tracking, graduation, or promotion of students.

The congressional charges to NAGB and to the NRC were constrained by P.L. 105-78 requirements that "no funds . . . may be used to field test, pilot test, administer or distribute in any way, any national tests" and that the NRC report be delivered by September 1, 1998.

The plan for pilot testing in March 1999 required that a large pool of potential VNT items be developed, reviewed, and approved by late fall of 1998, in order to provide time for the construction, publication, and distribution of multiple draft test forms for the pilot test. Given the March 1998 start-up date, NAGB, its prime contractor (AIR), and the subcontractors for reading and mathematics test development (Riverside Publishing and Harcourt-Brace Educational Measurement, respectively) faced a daunting and compressed schedule for test design and development.

A year after Congress placed restrictions on VNT development, it again considered issues relating to national testing. The Omnibus Consolidated Appropriations Act for fiscal 1999 (which emerged from negotiations between the White House and Congress in fall 1998) contained two related VNT components. The first set of provisions created a new section 447 of the General Education Provisions Act (GEPA), which added an additional restriction regarding the VNT:

• No funds can be used for pilot testing or field testing of any "federally sponsored national test . . . that is not specifically and explicitly provided for in authorizing legislation enacted into law." (There is currently no explicit authority for individualized national tests.)

The second set of provisions included the following requirements:

• NAGB shall continue to have exclusive authority over the direction and all policies and guidelines for developing voluntary national tests
• NAGB will report on three important VNT issues:
  (a) The purpose and intended use of the proposed tests;
  (b) A definition of the term "voluntary" as it pertains to the administration of the tests;
  (c) A description of the achievement levels and reporting methods to be used in reporting the test results.
• NAGB will report on its response to the National Research Council report (1999a) that evaluated NAEP, which repeated the criticism in some earlier evaluations that the process for setting achievement levels was "fundamentally flawed."
• The National Academy of Sciences (through the NRC) shall conduct a study regarding the technical feasibility, validity, and reliability of including test items from NAEP for 4th-grade reading and 8th-grade mathematics or from other tests in state and district assessments for the purpose of providing a common measure of individual student performance.

NAGB developed a work plan for 1999 that includes several important test development activities (based on Guerra, 1998):

(1) Detailed specifications describing the content and format of the reading and mathematics tests will be published.
(2) A specifications summary will be prepared and distributed.

(3) Both specifications versions will include sample items and will be available on the Internet.
(4) Efforts will continue to improve the pool of items that have already been written for the reading and math exams. Some additional questions may be written to be sure the proposed tests match NAEP in the range and distribution of item difficulty. Items will be reviewed to make sure they provide the strongest link possible with the NAEP achievement levels.
(5) NAGB will conduct an extensive series of focus groups and public hearings for its report on the purpose and use of the VNT and in defining the term "voluntary." Concurrently, NAGB will also deal with questions on how detailed any rules it makes should be and what issues should be left to state and local decision making.
(6) NAGB will continue its work on the issues of inclusion and accommodations for students with disabilities and limited English proficiency.
(7) All of these reports are due "not later than September 30, 1999." However, NAGB's executive committee recommended that the reports be submitted by June 30, 1999, to provide time for the reports to be considered in the deliberations on the future of the VNT during the upcoming session of Congress.

Figure 1-1 shows the timeline for key VNT development and test dates.

YEAR 1 EVALUATION

To carry out the original congressional mandate for an evaluation of VNT development efforts, the NRC appointed co-principal investigators to be assisted by several NRC staff members. After reviewing item development plans and examining item status and quality, the NRC issued an interim letter report (National Research Council, 1998a). The report expressed concern that the item development and review process was overly compressed, and it offered suggestions for rearranging the review schedule that were subsequently adopted by NAGB. The interim report also suggested the need to match VNT items to the descriptions of the NAEP achievement levels that would be used in reporting results.

The complete activities and results of the VNT year 1 evaluation were described in a final report issued on September 30, 1998 (National Research Council, 1999b). As described in that report, the primary focus of the year 1 evaluation was on the technical adequacy and quality of the development, administration, scoring, reporting, and use of the VNT that would aid test developers and policy makers at the federal, state, and local levels. The report covered specifications for the 4th-grade reading and 8th-grade mathematics tests; the development and review of items for the tests; and plans for subsequent test development activities. The last topic included plans for the pilot and field tests, for inclusion and accommodation of students with disabilities and for English-language learners, and for scoring and reporting the tests. The rest of this section summarizes the findings and conclusions of that report.

Test Specifications

The NRC found that the VNT test specifications were appropriately based on NAEP frameworks and specifications, but incomplete. The close correspondence with NAEP built on NAEP efforts to achieve a consensus on important reading and mathematics knowledge and skills and to maximize the prospects for linking VNT scores to NAEP achievement levels. However, the test specifications lacked information on test difficulty and accuracy targets and were not yet sufficiently tied to the achievement-level descriptions that will be used in reporting. Some potential users also questioned the

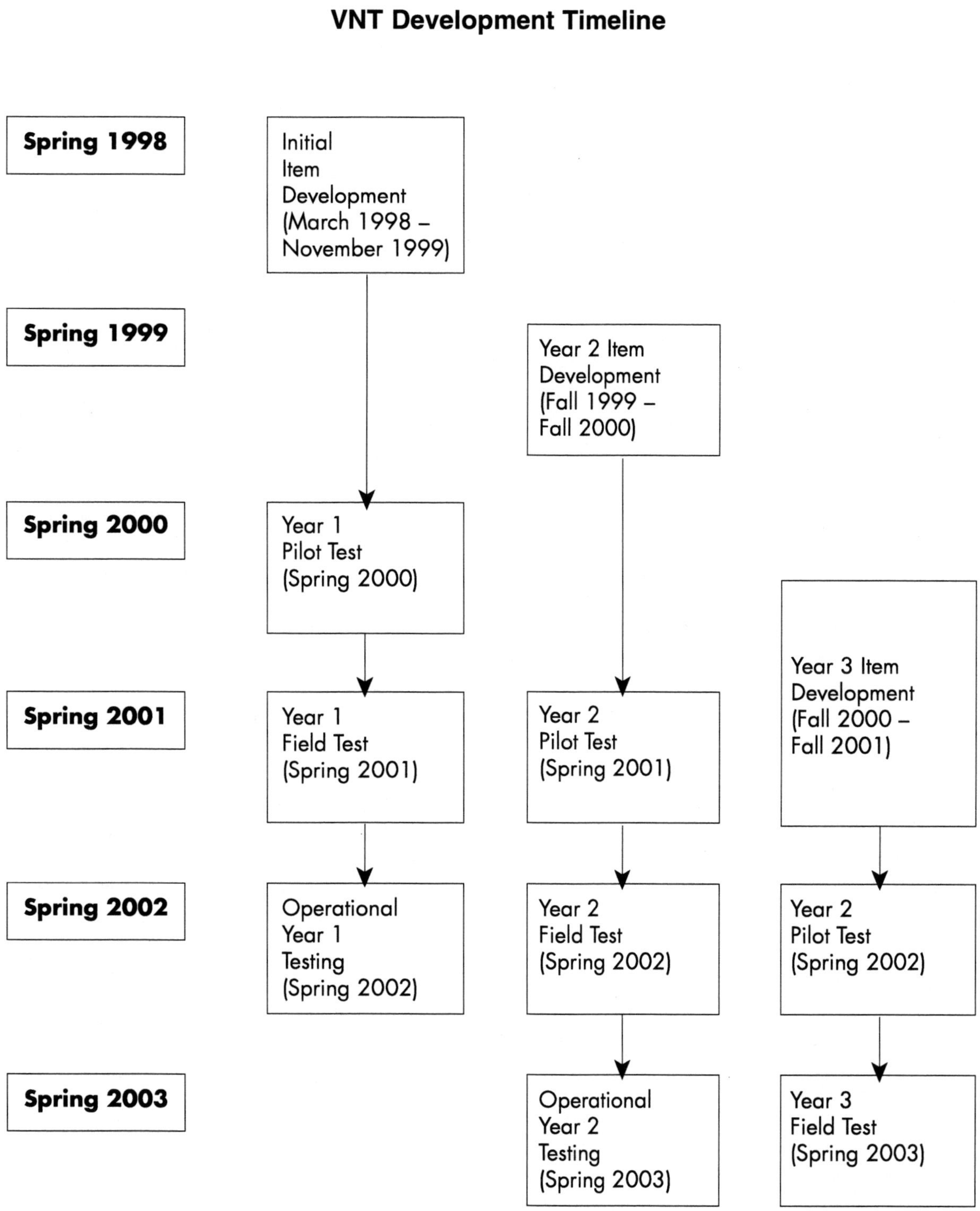

FIGURE 1.1 VNT development timeline.

decision to test only in English. The report recommended that test difficulty and accuracy targets and additional information on the NAEP achievement-level descriptions be added to the test specifications and that NAGB work to build a greater consensus for the test specifications to maximize participation by school districts and states.

Test Items

Because of significant time pressures, several item review and revision steps in 1998 were conducted simultaneously; as a result, opportunities were missed to incorporate feedback from them. Yet in terms of professional and scientific standards of test construction, the NRC concluded that the development of VNT items to date was satisfactory. NAGB and its consortium of contractors and subcontractors had made good progress toward the goal of developing a VNT item pool of adequate size and of known, high quality. Although it could not be determined whether that goal would be met, the procedures and plans for item development and evaluation were sound. At the same time, NAGB was urged to allow more time for future test development cycles so that the different review activities could be performed sequentially rather than simultaneously. The report also recommended that NAGB and its contractor develop a more automated item-tracking system in order to receive timely information on survival rates and the need for additional items. It said that item development should be tracked by content and format categories and by link to achievement-level descriptions so that shortages of any particular type of item could be quickly identified.

Pilot and Field Test Plans

The report concluded that the pilot and field test plans appeared generally sound with respect to the number of items and forms to be included and the composition and size of the school and student samples. It also concluded that more detail on plans for data analysis was needed and that some aspects of the design, such as the use of hybrid forms, appeared unnecessarily complex. It recommended that NAGB and its contractor develop more specific plans for the analysis and use of both the pilot and field test data, to include decision rules for item screening and accuracy targets for item parameter estimates, test equating, and linking. The report also recommended that greater justification be supplied for some aspects of the design plans, such as the use of hybrid forms, or that specific complexities be eliminated. NAGB was urged to prepare back-up plans in case item survival rates following the pilot test are significantly lower than anticipated.

Inclusion and Accommodation

The NRC found that plans for including and accommodating students with disabilities and English-language learners were sketchy and did not break new ground with respect to maximizing the degree of inclusion and the validity of scores for all students. Accommodation issues were not considered as an integral part of item development, and there were no clear plans for assessing the validity of accommodated scores. NAGB was urged to accelerate its plans and schedule for inclusion and accommodation of students with disabilities and limited English proficiency to increase both the participation of those student populations and to increase the comparability of VNT performance among student populations.

Reporting Plans

There were a number of potential issues in the reporting of test results to parents, students, and teachers that the NRC recommended be resolved as soon as possible, including: the adequacy of VNT items for reporting in relation to the NAEP achievement-level descriptions; mechanisms for communicating uncertainty in the results; and ways to accurately aggregate scores. The report also questioned whether and how additional information might be provided to parents, students, and teachers for students found to be in the "below basic" category.

The report recommended that NAGB accelerate its specification of procedures for reporting because reporting goals should drive most other aspects of test development. It said specific consideration should be given to whether and how specific test items will be linked and used to illustrate the achievement-level descriptions. It further recommended that attention be given to how measurement error and other sources of variation will be communicated to users, how scores will be aggregated, and whether information beyond achievement-level categories can be provided, particularly for students below the basic level of achievement.

OVERVIEW OF PLANNED YEAR 2 EVALUATION

In fall 1998 the Department of Education asked the NRC to continue its evaluation efforts of VNT development during fiscal 1999, with an expanded scope of work given the nature and timing of the test development process. In addition to a continued principal focus on the quality of the items being developed for the VNT, the year 2 committee considered the tests' purpose and questions of how the VNT would be used. Different test uses involve assumptions about inputs and consequences and suggest different technical constraints. The committee sought to identify important technical implications of proposed test use plans and to suggest how test use assumptions might be evaluated. Pertinent to this assessment is NAGB's charge to define the meaning of "voluntary," which affected the committee's analysis and recommendations about test use. We note that the "high stakes" issue in terms of VNT use for decisions about tracking, promotion, or graduation has already been considered in another congressionally mandated study of the VNT, which concluded that the VNT should not be used for such decisions (National Research Council, 1999d). However, we also considered this issue as part of the year 2 evaluation.

Scope of Work

The scope for year 2 of the VNT evaluation included analysis and comment on four issues, with several parts to three of them:

1. Item Quality

- Are the items developed for the VNT pilot test valid and informative measures for the test content? Are they free from obvious defects and not biased against ethnic or gender groups?
- Are the item development and review procedures used by NAGB and its contractor as complete and efficient as possible?
- Did the cognitive laboratory tryouts significantly improve item quality?

2. Technical Issues in Test Development

- Would the design for pilot testing result in items that represent the content and achievement-level specifications, are free of bias, and support test form assembly?
- Are the plans for assembling forms likely to yield highly parallel forms with adequate accuracy in classifying student results by achievement level?
- Are the revised designs for field testing and equating new forms and linking scores to NAEP achievement levels technically sound?

3. Inclusion and Accommodation

- Are the plans for including, accommodating, and meaningfully reporting results for students with special language and learning needs adequate and appropriate?

4. VNT Purposes and Practices

- Are NAGB's proposed rationales for test use clear and compelling? Are the assumptions behind the proposals sound?
- Are there potential unintended consequences of the proposed uses that should be considered?
- Do the plans for reporting VNT results make them accessible and meaningful for intended audiences?
- Are the plans for administration and governance feasible and free from unwarranted assumptions?

The committee examined these issues by reviewing and analyzing VNT procedures and products; soliciting expert review and analysis; reviewing the cognitive laboratory materials; collecting and analyzing field data; holding discussions with the relevant constituencies; and holding several information-gathering workshops.

The committee's first workshop, in February 1999, focused on test development and test designs. Committee members and other experts met with NAGB and AIR personnel to review and discuss:

- the extent to which information from content and bias reviews was used to cull and refine items and rubrics;
- findings from the analysis of results from the cognitive laboratories;
- implications of the VNT content reviews and achievement level reviews;
- current plans for pilot testing, field testing, and forms assembly;
- current plans for test linking and equating;
- current plans for inclusion and accommodation; and
- NAGB's December 23, 1998, reports to Congress on methods to be used to explore the purpose and definition of the VNT and plans for responding to the NRC's evaluation of NAEP achievement levels.

The second workshop, in April 1999, explored item development issues. Committee members met with other reading and mathematics content experts to review and discuss the extent to which item quality has improved through the original and more recent review and revision activities. The goals of this workshop were to evaluate the degree to which the items measure what the test developers state they are designed to measure, to assess specific problems that have appeared, and to match items to achievement levels to ascertain degrees of convergence. The workshop was held in closed session because of the need to review secure test items. Results of this review constitute a major component of this report and are presented in Section 3.

The third workshop, held in July 1999, reviewed and discussed:

- NAGB's June report to Congress on the purposes and intended uses of the VNT and the proposed definition of the term "voluntary;"
- the likelihood that intended decisions could be supported by the planned reporting metrics;
- NAGB's June report to Congress on the achievement levels and other reporting plans for the VNT; and
- plans for accommodating and reporting results for students who are English-language learners or who have special learning needs.

Interim Report

The committee issued an interim report earlier this year (National Research Council, 1999c) in light of NAGB's June 30, 1999, report to Congress and the White House so that the two reports, collectively, might contribute to the executive and legislative planning and decision making about the current status and potential future of the VNT. This report repeats virtually all of the material in the committee's interim report, with the findings and recommendations updated, modified, deleted, or expanded on the basis of new evidence and information about the continuing development of the VNT.

Report Purpose and Organization

The purpose of this final report is to provide evaluative information about the VNT, in the form of committee conclusions and recommendations, that will inform congressional and White House debate and discussions about the current status, issues, and future developments regarding the Voluntary National Tests.

The remaining sections of this report are organized around the primary issue areas in the committee's charge. Section 2 provides background on the purpose and use of the VNT, and it considers item (7) in our charge, the use of test items for tracking, promotion, or graduation; it also considers broad issues of VNT purposes and practices. Section 3 discusses the quality of items that are now ready for pilot testing and the likelihood that a sufficient number of additional items will be developed to an appropriate level of quality in time for inclusion in a spring 2000 pilot test. It addresses items (1), (2), (4), and (5) in our charge. Section 4 discusses pilot and field test design issues, item (3) in our charge. Section 5 considers issues on the inclusion and accommodation of students in the VNT with special language and learning needs, item (6) in our charge. Section 6 discusses the topic of VNT reporting. Finally, Section 7 presents the committee's overall conclusions and lists all our recommendations.

2

Purpose and Use

In its statement of purpose, released on June 23, 1999, the National Assessment Governing Board (1999e:1) responded to the congressional request that it:

> Determine and clearly articulate the purpose and intended use of any proposed federally sponsored national test. Such report shall also include—(A) a definition of the meaning of the term 'voluntary' in regards to the administration of any national test; and (B) a description of the achievement levels and reporting methods to be used in grading any national test.

Although NAGB has had exclusive authority to oversee the policies, directions, and guidelines for developing the VNT since November 1997, this report was the board's first general statement about the overall purposes of the Voluntary National Tests. Thus, we carefully reviewed the process leading to the report, as well as its content. To accomplish this, we reviewed the following documents:

- The Voluntary National Test: Purpose, Intended Use, Definition of Voluntary and Reporting (National Assessment Governing Board, 1999e)
- Overview: Determining the Purpose, Intended Use, Definition of the Term Voluntary, and Reporting for the Proposed Voluntary National Test (National Assessment Governing Board, 1999a)
- Public Hearings and Written Testimony on the Purpose, Intended Use, Definition of the Term "Voluntary," and Reporting of the Proposed Voluntary National Test. Synthesis Report. April (National Assessment Governing Board, 1999d)
- *High Stakes: Testing for Tracking, Promotion, and Graduation* (National Research Council, 1999d)
- The National Assessment Governing Board and Voluntary National Tests: A Status Report (Guerra, 1998).

THE NAGB PROCESS AND REPORT

Two Scenarios

The National Assessment Governing Board did not take a position for or against the VNT, but simply responded to the conditions of the congressional request. By March 1999 NAGB had prepared a set of background materials that offered two competing scenarios for the purpose and use of the VNT: a "public policy model" and an "individual decision model" (National Assessment Governing Board, 1999e:Appendix); see Appendix A for NAGB's descriptions of the two scenarios. Congress set a deadline of September 30, 1999, for NAGB's report, but the board responded 3 months early in order to give Congress sufficient time to use the report in its deliberations about the future of the VNT.

The public policy and individual decision models had several elements in common:

- overall purpose;
- no federal requirement, sanction, or reward for participation;
- no federal receipt, scoring, tabulation, or reporting of VNT data;
- extensive feedback of individual performance data with results reported in terms of NAEP achievement levels;
- no ties between the VNT and a specific curriculum, teaching method, or approach;
- no intent to diagnose specific learning problems or English language proficiency;
- no use as the sole criterion in high stakes decisions about individual students; and
- no evaluation of instructional practices or programs, or of school effectiveness.

In addition, some issues were left open in NAGB's scenarios for decisions under either policy model: whether (after development by NAGB) the VNT would be under the operational control of a federal agency or would be marketed by one or more commercial test publishers and who would pay for the VNT after its initial development and administration (which is to be funded by the federal government).

Distinctions between the public policy model and the individual decision model included the definition of "voluntary," the intended use of the tests, some details of the reporting plan, and several aspects of test administration. Under the public policy model, public or private school authorities would "volunteer" on behalf of their students. Depending on state and local law and policy, public schools might volunteer at the state or district level. Parents might "opt out" of testing if permitted by public or private school authorities. In contrast, under the individual decision model, parents (alone) would decide whether a student participated in the VNT.

Under the public policy model, parents, students, and authorized educators would receive reports of VNT performance; under the individual decision model, only parents and students would receive reports. Some norm-referenced information[1] (at the national level) would supplement achievement-level reporting under either model, but under the individual decision model, there would be no comparisons at the class, school, district, or state level. In addition, under the public policy model, individual data could be compiled by state or local participants, subject to technical guidance on reporting from a central source. Such institutional participants "will bear responsibility for using resulting data in valid, appropriate ways" (National Assessment Governing Board, 1999e:Appendix).

[1] A norm-referenced test is used to ascertain an individual's status with respect to the performance of other individuals on the test (Popham, 1990:26).

Use of VNT results by persons other than students and parents could be far more extensive under the public policy model than in the individual decision model. In the individual model, "follow-up with school/teacher is up to the parent (National Assessment Governing Board, 1999e:Appendix)." In the public policy model, parent and teacher follow-up would be recommended but determined at the state, district, or school level; results could also be used to make decisions about individual students, or they could be compared with student performance on state or local tests. Educational authorities might wish to use test results as "an external anchor to their state tests" (National Assessment Governing Board, 1999e:Appendix) or, possibly, as part of school accountability systems.

Finally, there would be a number of operational differences between the testing programs under the public policy model and the individual decision model, for example, in dissemination strategies, training of relevant school personnel, and systems for reporting and responding to parental queries.

Public Comment Process

NAGB's descriptions of possible purposes, uses, and implementation strategies for the VNT provided a framework for public comment in a series of open hearings. More than 2,000 invitations were issued for hearings in Atlanta, GA, Washington, DC, Chicago, IL, and San Francisco, CA, from March 29 to April 12, 1999; 51 people and organizational representatives provided oral or written testimony in response to the invitations. About two-thirds of responses came from school systems or educational organizations, while the remainder came from research organizations, testing companies, corporations, foundations, or other national organizations (National Assessment Governing Board, 1999d). After these hearings, NAGB prepared a draft report for review at its May 1999 meeting, and the draft was released for comment by mail, on NAGB's Internet web site and in the *Federal Register*. A final hearing with state and district testing officers was held on June 12, 1999, at the annual Large Scale Assessment Conference of the Council of Chief State School Officers, and a revised report was approved at a special meeting of the board on June 23, 1999. Although the weak public response to NAGB's alternative scenarios was disappointing—and possibly indicative of a decline in public interest in the VNT—the committee finds that the report reflects an effort by NAGB to bring a wide range of views into its deliberations in a limited time frame.

NAGB Proposals

NAGB's report responds to each of the congressional charges: purpose, intended use, definition of voluntary, and reporting issues. In addition, it comments on other issues of implementation and on the linkage of the VNT to the NAEP. The board suggests to Congress and the President that the following statement of purpose be considered (National Assessment Governing Board, 1999e:5):

> To measure individual student achievement in 4th grade reading and 8th grade mathematics, based on the content and rigorous performance standards of the National Assessment of Educational Progress (NAEP), as set by the National Assessment Governing Board (NAGB).

In recommending this statement of purpose, NAGB pointed to the importance of grades 4, 8, and 12 in schooling and to the NAEP's focus on achievement at those grade levels. Proficiency in reading by grade 4 and in mathematics by grade 8 is generally regarded as fundamental to academic success. Since the NAEP provides scores only for population groups—nationally or for states—the VNT would be the first national test to provide individual scores consistent with the NAEP frameworks and achievement levels. However, NAGB noted its sensitivity to fears of federal encroachment on state and local control of schooling, including the emergence of a national curriculum. The report thus argues for "checks and balances in the governance and operation of the voluntary national testing program to address these reasonable concerns" (National Assessment Governing Board, 1999e:6).

Concern with federal intrusion is strongly expressed in NAGB's definition of voluntary in the Voluntary National Tests (National Assessment Governing Board, 1999e:7):

> The federal government shall not require participation by any state, school district, public or private school, organization, or individual in voluntary national tests. The federal government also shall not make participation in voluntary national tests a specified condition for receiving federal funds or require participants to report voluntary national test results to the federal government.

Because the federal role in determining who will participate is so limited by this definition, the report stated that the definition of voluntary must accommodate "a wide range of diversity of governance authority," and it suggests an array of levels at which the decision to participate in the VNT could be made, consistent with state and local law and policy (National Assessment Governing Board, 1999e:8):

> Public and private school authorities should be afforded the option to participate in the voluntary national tests. For public schools, state and/or local law and policy should determine whether the initial decision to participate is made at the state level or at the local district level. Where state law or policy provides that the initial decision be made at the state level, and the state decides not to participate, school districts should be afforded the opportunity to decide whether to participate, to the extent permitted by state and local law and policy.
>
> For private schools, the decision to participate should be made by the appropriate governing authority.
>
> Parents may have their children excused from testing as determined by state and local law and policy in the case of public schools. In the case of private schools, parents may have their children excused from testing as determined by the policy of the appropriate governing authority.
>
> Parents whose schools are not participating but want their children to take the voluntary national tests should have access to the tests either through a qualified individual or testing organization before the tests are released to the public or through dissemination procedures at no or minimal cost (e.g., public libraries and the Internet) after the tests are released to the public.

Clearly, this definition spreads responsibility for test-taking broadly across potentially interested parties—but it explicitly excludes federal authorities. States, school districts, private schools, or parents could all opt for the VNT, while parents could opt out, when permitted by local policy.

The report suggests to Congress and the President that they consider the following statement of intended use of the Voluntary National Tests (National Assessment Governing Board, 1999e:9):

> To provide information to parents, students, and authorized educators about the achievement of the individual student in relation to the content and the rigorous performance standards for the National Assessment, as set by the National Assessment Governing Board for 4th grade reading and 8th grade mathematics.

In its rationale, NAGB focuses on the measurement of individual student achievement in relation to NAEP performance standards, consistent with its preceding statement of purpose, and it identifies the audience for VNT results as "parents, students, and authorized educators." The report suggests several ways in which the VNT could affect future educational progress, for example (National Assessment Governing Board, 1999e:10):

> (1) parents could become more involved with the child's education, (2) students could study hard and learn more, (3) teachers could work more to emphasize important skills and knowledge in the subjects tested without narrowing or limiting their curricula, and (4) parents, students, and teachers could have a means for better communication about the child's achievement.

However, NAGB recognizes that the achievement of such outcomes would depend "on local effort, resources, skill, and persistence" (National Assessment Governing Board, 1999e:10) and thus would not automatically follow from the VNT or any other testing program.

The report also noted that VNT results would probably be used in appropriate ways other than its enumerated uses of individual test scores, "at the discretion of the participating state, district, or private school authorities, who would be responsible for following appropriate technical standards and validation procedures" (National Assessment Governing Board, 1999e:10). The report notes that the VNT, like NAEP, does not reflect an approved curriculum, instructional method, or approach; that it is not designed "to diagnose specific learning problems or English language proficiency;" and that—consistent with professional testing standards (citing standards of the American Educational Research Association et al. [in press] and the report of the National Research Council [1999c]—the VNT "should not be used as the sole criterion in making high stakes decisions (e.g., placement or promotion) about individual students" (National Assessment Governing Board, 1999e:11).

However, NAGB did not accept the recommendation of the National Research Council (1999c) that the VNT not be used for any high stakes decision about individual students. Rather, the report explains (National Assessment Governing Board, 1999e:11):

> It is conceivable that circumstances could exist in which state and/or school district authorities would find information from the voluntary national test useful, in concert with other valid information, in making "high stakes" decisions about individual students. Therefore, it seems prudent not to foreclose such use absolutely. The Governing Board recommends that participating states, districts, and schools should be afforded the discretion to decide whether to use the voluntary national test for high stakes purposes and the responsibility for ascertaining that such uses are valid.

NAGB recommends extensive feedback of information about individual student performance and about the test itself in the context of the performance standards (achievement levels of basic, proficient, and advanced) determined by NAGB (National Assessment Governing Board, 1999e:11):

> Consistent with the purpose and intended use of the voluntary national tests, the National Assessment Governing Board suggests that results of the voluntary national tests be provided separately for each student. Parents, students, and authorized educators (those with direct responsibility for the education of the student) should receive the test results report for the student. Test results for the student should be reported according to the performance standards for the National Assessment of Educational Progress (NAEP). These are the NAEP achievement levels: Basic, Proficient, and Advanced. All test questions, student answers, and an answer key should be returned with the test results; it will be clear which questions were answered correctly and which were not. The achievement levels should be explained and illustrated in light of the questions on the test. Also, based on the nationally representative sample of students who participated in the national tryout of the test the year before, the percent of students nationally at each achievement level should be provided with the report.

The report recognizes a likely demand for aggregate reporting of test results, "for the nation as a whole or by state, district, school, or classroom," but it states that no compilations should be provided automatically by the VNT program, "since the purpose and use of the testing program are directed at individual student level results" (National Assessment Governing Board, 1999e:11). Thus, participants who want aggregate results "should be permitted to obtain and compile the data at their own cost, but they will bear the full responsibility for using the data in appropriate ways and for ensuring that the uses they make of the data are valid" (National Assessment Governing Board, 1999e:11). NAGB states that it would develop and provide guidelines for "appropriate and valid" compilation and reporting of VNT results and that its guidelines would "explicitly require full and clear disclosure about exclusions and/or absences from testing, so that results and comparisons would be accurately portrayed" (National Assessment Governing Board, 1999e:12).

ASSESSMENT AND RECOMMENDATIONS

Individuals who are already familiar with the proposal for Voluntary National Tests will find little that is new in NAGB's report on purpose, intended use, meaning of the term voluntary, and reporting. Rather, it codifies several elements of the proposal as initially offered by the Clinton administration and modified when NAGB assumed control of test development:

- a focus on individual performance in reading in the 4th grade and mathematics in the 8th grade;
- the effort to link VNT content, standards, and reporting to the National Assessment of Educational Progress;
- extensive feedback of test results to individual students;
- voluntary participation by states or local public and private school authorities; and
- a clearly defined prohibition of federal participation in the VNT program, beyond its support of test development and, possibly, operational costs.

The NAGB report also articulates two significant modifications of the VNT design: permitting individual parents to opt for the VNT program, even if their children's schools do not participate, and use of norm-referenced, as well as criterion-referenced,[2] score reports to individual students.

The committee finds one significant element lacking, however, both in the original proposal for the VNT and in the NAGB report: evidence to support the belief that the VNT, if implemented, would in fact have favorable effects. The proposal for the Voluntary National Tests, as articulated by the Clinton administration and revised by the National Assessment Governing Board, *assumes* positive educational effects of individual achievement testing based on high standards, as embodied in NAEP. President Clinton argued that "good tests will show us who needs help, what changes in teaching to make, and which schools need to improve" (National Research Council, 1999e). Among other potential benefits suggested in early discussion of the VNT was the belief that information provided by the VNT would lead to greater parental involvement—a positive outcome, given that the relationship between parental involvement and student achievement is one of the more consistent findings of educational research. Yet critics have argued that information from the VNT would simply perpetuate beliefs of failing students and their parents about the students' inability to learn and might not lead to any improvement in educational processes at all.

The NAGB report on VNT purpose and use takes an important first step in delineating what the VNT is being designed to do: "to measure individual student achievement in 4th-grade reading and 8th grade mathematics, based on the content and rigorous performance standards of the National Assessment of Educational Progress" (National Assessment Governing Board, 1999e:5). The report includes a vignette illustrating possible outcomes from implementation of the program and then states: "The story emphasizes that, while having widely recognized standards and assessments can provide focus for planning and a common language for students, parents, and teachers, what is most important is what parents, students, and educators actually do with that knowledge." Much more dialogue is needed, however, to articulate the range of ways in which the information provided by the VNT might or might not lead to positive changes.

---

[2] A criterion-referenced test is used to ascertain an individual's status with respect to a defined assessment domain. Whereas a norm-referenced test references performance to a norm group, a criterion-referenced test references performance to a defined set of criterion behaviors, such as a specific type of reading skill (i.e., the ability to infer the main idea of a reading passage) or a particular mathematics skill (i.e., the ability to solve word problems based on two or more arithmetic operations) (Popham, 1990:27).

The revised test standards (American Educational Research Association et al., 1999), just recently approved, place greater emphasis on the need to consider consequences of test use. In this regard, AIR has proposed an evaluation of the VNT on instruction, parental involvement, and student motivation to accompany implementation of the VNT (American Institutes for Research, 1999e). The committee supports the idea of studies of the consequences of the VNT, based in part on evidence from research on the consequences of other similar assessments (see also National Research Council, 1999c). In this regard, the committee believes that a good starting point for consideration of potential effects is from existing research on the impact of state assessments, many of which use standards-based reporting and bear other similarities—in design and purpose—to the proposed VNT.

The committee believes high priority should be given to articulation of the benefits to be derived from the VNT, as well as to a search for evidence of the likelihood that such benefits would be realized. That evidence may be found through a review of similar programs of educational initiatives and in continuing evaluation of the VNT, if and when implemented.

RECOMMENDATION 2.1 High priority should be given to the articulation of the potential educational effects of the VNT and to the development of a program of research and evaluation to determine whether and how the VNT contributes to improved educational outcomes.

On the whole, the recommendations of the National Assessment Governing Board follow its public policy model, rather than the individual decision model. That is, students may be "volunteered" by state or local school authorities or by private schools. Score reports are to go to "authorized educators," as well as to students and their parents. Institutional participants have the option of aggregating data, subject to guidelines to be developed later. Thus, it may be possible to report and compare scores at the classroom, school district, or state levels, which would not be possible under the individual decision model.

In addition to the intended use, as described above, NAGB does not limit or proscribe other uses of VNT scores by institutional participants (National Assessment Governing Board, 1999e:10):

> The Governing Board does not assume that uses of data from voluntary national tests beyond the intended use described above are necessarily inappropriate or should be prohibited to states, districts, and private schools. Any such additional use of voluntary national test data would be done at the discretion of the participating state, district, or private school authorities, who would be responsible for following appropriate technical standards and validation procedures.

In particular, NAGB states:

> The voluntary national tests are not intended to be used as the sole criterion in making "high stakes" decisions about individual students. . . . The primary consideration is that, when making decisions on matters such as promotion, retention, or placement, scores from large-scale assessments should only be used in combination with other information about student achievement.

This recommendation is consistent with professional standards for test use and with the NRC report on high-stakes testing (National Research Council, 1999d); however, the Board proposes no mechanism for enforcing its recommendation.

The NAGB report also ignores several other standards of appropriate test use when it adds, "it is conceivable that circumstances could exist in which state and/or school district authorities would find information from the voluntary national test useful, in concert with other valid information, in making 'high stakes' decisions about individual students" (National Assessment Governing Board, 1999e:11). NAGB's report does not suggest what those circumstances might be. The NRC report enumerated an extensive list of the circumstances under which the VNT, as proposed, might be used to make high stakes decisions about individuals, and it found that in each case use of the VNT would be invalid or otherwise unethical (National Research Council 1999c:Ch. 10). For example, there is no provision—such as an alternative form—for students to retake the VNT, and rapid public release of test items and answers would surely preclude retaking the then-current form of the test. Yet the opportunity to retake a high-stakes examination is one of the basic criteria of appropriate test use in decisions about individuals. In addition, the explicit and deliberate absence of alignment of the VNT (or its model, NAEP) with existing curricula would appear to invalidate its use either as a criterion for promotion or as a guide to placement. The committee is left without any information about the circumstances under which NAGB envisions that state or school district authorities could appropriately use VNT information to make high-stakes decisions about individual students. The NAGB report is also silent on what would be done if the VNT were used inappropriately to make high-stakes decisions.

NAGB adopted its public policy model (slightly modified) with respect to participation in the VNT and uses of the resulting information. Yet the focus of the VNT—both in terms of purpose and intended use—is on the achievement of the individual student. The proposed approach is problematic because of its effort to combine individual and public action. It attempts to inform, but does not protect or empower parents or students. Consider, in contrast, the role of the primary institutional participants—state, local, and private school authorities. In NAGB's report, they are insulated from federal intrusions and from tangible responsibility for appropriate use of information from the VNT in decisions about individual students, despite NAGB's expressed concern about some potential misuses of the tests. In effect, NAGB (or its operational successor) would be no more responsible for appropriate use of the VNT than any commercial publisher is for any existing test. Although there are instances in which findings from the National Assessment of Educational Progress (the other major testing program for which NAGB is responsible) have been misused, the potential for serious misuse is much greater in the case of the VNT, with potential significant consequences for individual students, teachers, and schools. Thus, NAGB should rethink its positions with respect to prevention and control of potentially inappropriate uses of the VNT.

RECOMMENDATION 2.2 The National Assessment Governing Board should develop explicit and detailed guidelines, practices, and enforcement mechanisms for the appropriate use of the Voluntary National Tests relative to high-stakes decisions about individual students or about teachers, classrooms, schools, or other educational units. Those guidelines should illustrate uses of the VNT relative to high-stakes decisions that are inappropriate and explicitly state the potential consequences of such inappropriate uses.

For example, one of the recommendations of the 1999 NRC report on appropriate test use (National Research Council, 1999c) was that test producers adopt "truth in labeling," much as food and pharmaceutical producers are now required to do. NAGB has asked its prime contractor, the American Institutes for Research, to develop and test materials for reporting scores and other information about the VNT to students and parents. However, none of the materials for parents or teachers that have been presented to the committee bear on inappropriate uses of the test information or on what parents or teachers can do to prevent misuse of the tests.

Just as NAGB provides school authorities with broad license to use information from the VNT to make decisions about individual students, it also provides a similar license for the use of a variety of aggregations of test scores—at the classroom, school, district, or state level. While NAGB does not plan for the automatic aggregation of VNT score data, it accepts that many participants (e.g., schools, districts, states) may wish to aggregate results. The NAGB report states that it "would develop and provide guidelines and criteria for use by states, districts, and schools for compiling and reporting the data from the voluntary national tests in ways that are appropriate and valid" (National Assessment Governing Board, 1999e:11), and it insists that the users could be responsible for the appropriate uses of aggregate data. However, the committee has not seen a framework or outline for such guidelines. Apart from requiring "full and clear disclosure about exclusions and/or absences from testing" (National Assessment Governing Board, 1999e:12), the report does not indicate concern about potential misuses of aggregate data—such as those pertaining to evaluation of "instructional practices, programs, or school effectiveness"—that were noted in its draft VNT scenarios. Again, the report provides no indication of what NAGB would do in the event of misuse of aggregated VNT information. The committee believes this is a major deficit in the NAGB report:

RECOMMENDATION 2.3 The National Assessment Governing Board should develop explicit and detailed guidelines, practices, and enforcement mechanisms for the appropriate compilation and use of aggregate data from administrations of the Voluntary National Tests relative to high-stakes decisions about teachers, classrooms, schools, or other educational units.

Most of the NAGB report and the committee's assessment deal with some of the more technical and educational issues surrounding the VNT, yet there are a formidable number of practical issues that must be addressed in the near future. These include issues of funding, namely: Who will pay for the VNT and delivery, and who will administer, score, and report results? Will it be a federally funded program or turned over to the commercial sector for development, administration, and reporting? Will there be central delivery management and oversight or a free-market model with many firms licensed to carry out this function (see Guerra, 1998)? Will there be funding stages (from the federal government to "volunteers"), or will there be continued federal funding? Who will be responsible for test security? Who will provide technical support and answer questions about test use and test results? In developing proposals for delivery systems and funding, NAGB can determine the extent of national interest in participation under each of the scenarios it considers.

These and related issues need to be considered now because they have a direct bearing on potential participation and the timing of decisions about participation. There is a clear need for NAGB to begin to consider information gathering regarding potential test administration (i.e., operational) issues. For users to decide whether to participate, they will need to know about funding and operational delivery systems (e.g., procedures for administration, scoring, reporting, etc.). Congress will almost certainly want to know about the scope of VNT use.

RECOMMENDATION 2.4 The National Assessment Governing Board should continue to develop plans for how the VNT would operate. Specifically, it should develop proposals for operational delivery systems for the VNT and for funding ongoing development and delivery costs so that potential users can make decisions about their participation, based on the costs as well as the potential educational value of the VNT.

3

Item Quality and Readiness

The primary focus of this section is the extent to which the VNT test items are likely to provide useful information to parents, teachers, students, and others about whether students have mastered the knowledge and skills specified for basic, proficient, or advanced performance in 4th-grade reading and 8th-grade mathematics. The information provided by any set of items will be useful only if it is valid, meaning that the items measure the intended areas of knowledge and do not require extraneous knowledge or skills. In particular, test items should not require irrelevant knowledge or skills that might be more available to some ethnic, racial, or gender groups than to others: that is, they should not be biased. Test information also will be useful only if it is reliable, meaning that a student taking alternate forms of the test on different occasions is very likely to achieve the same result.

The committee's review of the quality of the VNT items thus addresses four of Congress' charges for our evaluation: (1) the technical quality of the items; (2) the validity, reliability, and adequacy of the items; (4) the degree to which the items provide valid and useful information to the public; and (5) whether the test items are free from racial, cultural, or gender bias. The NRC's Phase I report (National Research Council, 1999b) included only a very limited evaluation of item quality. No empirical data on item functioning were available, and, indeed, none of the more than 3,000 items that had been written had been through the contractor's entire developmental process or NAGB's review and approval process. Our review of items in relatively early stages of development suggested that considerable improvement was possible, and the contractor's plans called for procedures that made further improvements likely.

This review of VNT items initially addressed two general questions related to item quality:

1. Does it seem likely that a sufficient number of items will be completed in time for inclusion in a spring 2000 pilot test?
2. Are the completed items judged to be as good as they can be prior to the collection and analysis of pilot test data? Are they likely to provide valid and reliable information for parents and teachers about students' reading or math skills?

In addressing these questions, the committee was led to two additional questions relating to item quality:

3. Do the NAEP descriptions of performance for each achievement level provide a clear definition of the intended domains of test content?
4. How completely will the items selected for each test form cover the intended test content domains?

To answer these questions, the committee reviewed the following documents from NAGB and the prime contractor, American Institutes for Research (AIR):

- Reading and math test specification matrices (National Assessment Governing Board, 1998b; 1998c)
- Report on the Status of the Voluntary National Tests Item Pools (American Institutes for Research, 1999f)
- Flowchart of VNT New Item Production Process (American Institutes for Research, 1999d)
- VNT: Counts of Reading Passages Using Revised Taxonomies, June 24, 1999 (American Institutes for Research, 1999k)
- Final Report of the Study Group Investigating the Feasibility of Linking Scores on the Proposed VNT and NAEP (Cizek et al., 1999)
- VNT in Reading: Proposed Outline for the Expanded Version of the Test Specifications (American Institutes for Research, 1999n)
- VNT in Mathematics: Proposed Outline for the Expanded Version of the Test Specifications (American Institutes for Research, 1999m)
- Cognitive Lab Report: Lessons Learned (American Institutes for Research, 1999a)
- Training Materials for VNT Protocol Writing (American Institutes for Research, 1999j)
- VNT: Report on Scoring Rubric Development (American Institutes for Research, 1998o)
- Cognitive Lab Report (American Institutes for Research, 1998d)
- VNT Interviewer Training Manual (American Institutes for Research, 1999o)
- Technical Specifications, Revisions as of June 18, 1999 (American Institutes for Research,, 1999i)

In addition, committee and staff members examined item folders at the contractor's facility and reviewed information on item status provided by AIR in April. During our April meeting, committee members and a panel of additional reading and mathematics assessment experts reviewed and rated samples of 120 mathematics items and 90 reading items. Updated item status data, including more specific information on the new items being developed during 1999, were received in July and discussed at our July meeting. The committee's review of item quality did not include separate consideration of potential ethnic or gender bias. The contractor's process for bias review in year 1 was reviewed in the Phase I report (National Research Council, 1999b) and found to be satisfactory, and no new bias reviews have been conducted. (The committee does have suggestions in Chapter 4 for how pilot test data might be used in empirical tests of ethnic and gender bias.)

The remainder of this chapter describes the committee's review, findings, and recommendations relative to each of the four item quality questions listed above.

ITEM DEVELOPMENT

As noted above, the committee reviewed item development status at two different times in 1999. In April we received information on the status of items that were developed in prior years for use in selecting a sample of completed items for our review. In July we received updated information, including information on the new items written in 1999 to supplement the previous item pool.

Item Status as of April 1999

The VNT Phase I evaluation report suggested a need for better item tracking information. At our February 1999 workshop, the contractor presented plans for an improved item status tracking system (American Institutes for Research, 1999f). We subsequently met with NAGB and the contractor's staff to make arrangements for identifying and obtaining access to the items needed for our review. The contractor provided additional information on the item tracking database and a copy of key information in the database for our use in reviewing the overall status of item development and in selecting a specific sample of items for review. We also visited the contractor facilities and were allowed access to the system for storing hard-copy results of the item development and review for each item. We examined the item folders for a small sample of items and found that the information was generally easily found and well organized.

Our primary concern in examining the item status information was to determine how far along each item was in its development process and how far it had yet to go. We were interested in identifying a sample of completed items so that we could assess the quality of items that had been through all of the steps in the review process. We also wanted to assess whether it was likely that that there would be a sufficient number of completed items in each content and format category in time for a spring 2000 pilot test.

The contractor suggested that the most useful information about item status would be found in two key fields in the database for each item. The first field indicated whether consensus had been reached in matching the item to NAEP achievement levels: if this field was blank, the item had not been included in the achievement-level matching and was not close to being completed. The second field indicated whether the item had been through a "scorability review" and, if so, whether further edits were indicated. The scorability review is a separate step in the contractor's item development process that involves expert review of the scoring rubrics developed for open-ended items to identify potential ambiguities in the rules for assigning scores to them. A third key field was added to the database, at our request, to indicate whether or not the item had been reviewed and approved by NAGB's subject area committees.

The committee reviewed the revised database to determine the number of items at various levels of completeness for different categories of items. Table 3-1 shows levels of completeness for mathematics items by item format and content strand. Table 3-2 shows the same information for reading items, by stance and item format. As of April 1999, only one-sixth (16.6%) of the required mathematics items and one-eighth (12.3%) of the required reading items were completed. In addition, at least 161 new mathematics items were required to meet item targets for the pilot test. The contractor indicated that 200 new mathematics items were being developed in 1999; however, they could not, at that time, give us an exact breakdown of the number of new items targeted for each content and item format category.

For reading, the situation is more complicated. Current plans call for 72 passages to be included in the pilot test. Each passage will be included in two distinct pilot test forms with different sets of questions about the passages in each of the forms. This design will increase the probability that at least

TABLE 3-1 Mathematics Item Status (as of April 1999)

Item Format[a]	Content Strand	Needed for Pilot	Fully Ready	Awaiting NAGB Review	Awaiting Ach. Level Matching	In 1999 Cog Labs	Awaiting Scoring Edits	Total Items Written	Items Needed
ECR	Algebra and functions	18	1	0	0	0	6	7	11
	Geometry and spatial sense	18	0	1	0	3	4	8	10
	Other	None	1	1	0	5	13	20	0
	Subtotal	36	2	2	0	8	23	35	21
SCR/3 points	Algebra and functions	18	6	1	0	4	15	26	0
	Data analysis, statistics, and probability	18	1	5	0	11	8	25	0
	Geometry and spatial sense	18	0	2	0	8	16	26	0
	Measurement	18	8	10	1	13	9	41	0
	Number	36	7	10	1	11	14	43	0
	Subtotal	108	22	28	2	47	62	161	0
SCR/2 points	Algebra and functions	18	1	1	0	1	1	4	14
	Data analysis, statistics, and probability	18	0	6	0	2	1	9	9
	Geometry and spatial sense	18	2	4	0	4	7	17	1
	Measurement	None	2	4	0	4	1	11	0
	Number	18	1	2	0	3	1	7	11
	Subtotal	72	6	17	0	14	11	48	35
GR	Algebra and functions	None	1	7	1	1	0	10	0
	Data analysis, statistics, and probability	18	6	21	2	4	0	33	0
	Geometry and spatial sense	18	5	21	1	2	0	29	0
	Measurement	36	0	14	5	7	0	26	10
	Number	36	5	25	1	3	0	34	2
	Subtotal	108	17	88	10	17	0	132	12
MC	Algebra and functions	198	26	99	15	4	0	144	54
	Data analysis, statistics, and probability	108	11	71	1	4	0	87	21
	Geometry and spatial sense	126	38	64	1	5	0	108	18
	Measurement	126	11	137	1	8	0	157	0
	Number	198	46	222	1	13	0	282	0
	Subtotal	756	132	593	19	34	0	778	93
Total		1,080	179	728	31	120	96	1,154	161

[a]ECR = extended constructed response; SCR = short constructed response; GR = gridded; MC = multiple choice.

TABLE 3-2 Reading Item Status (as of April 1999)

Items	Needed for Pilot	Fully Ready	NAGB Review	Cognitive Labs	Scoring Rubric Edits	Total Written	New Items Needed[a]
By Stance							
Initial understanding	130	15	125	29	6	175	
Develop interpretation	572	77	597	62	42	778	
Reader-text connection	108	5	67	23	29	124	
Critical stance	270	36	219	33	27	315	
Subtotal	1,080	133	1,008	147	104	1,392	0
By Item Format[b]							
ECR	48	1	23	19	31	74	
SCR	192	20	150	53	55	278	
MC	840	112	835	75	18	1,040	
Subtotal	1,080	133	1,008	147	104	1,392	0

[a]See text and Table 3-3.
[b]ECR = extended constructed response; SCR = short constructed response; MC = multiple choice.

TABLE 3-3 Reading Passage Review Status (as of April 1999)

Passage Type	Completed NAGB Review		Completed NAGB and Edits		Needs More Items		Total Passages Written	Passage Length Issues[a]	Additional Passages Needed
	Both Sets	One Set	Both Sets	One Set	Both Sets	One Set			
Long literary	2	5	11	5	7	5	23	3	0
Medium literary	0	3	8	2	0	2	10	0	2
Short literary[b]	6	0	10	1	0	1	11	7	1
Medium information[c]	0	9	9	5	0	5	14	11	0
Short information	5	3	11	0	0	0	11	10	1
TOTAL	13	20	49	13	7	13	69	31	4

[a]The seven long literary passages needing more items for both sets appear to have been developed as medium literary passages.
[b]One short literary passage is too short (< 250 words) and six are between short and medium length. All of the short information passages with length problems are between 300 and 350 words, which is neither short nor medium. Two additional short information passages are classed as medium information due to length, but they have no pairing nor intertextual items.
[c]Medium information entries are passage pairs plus intertextual questions.

one set (or perhaps a composite of the two different sets) will survive item screening in the pilot test. As of April, there were no passages for which both item sets had completed the review and approval process. Table 3-3 shows the number of passages at each major stage of review and development, the number of passages for which additional items will be needed, and the number of additional passages that will be needed. One further issue in reading is that many of the passages have word counts that are outside the length limits indicated by the test specifications. In most cases, these discrepancies are not large, and NAGB may elect to expand the limits to accommodate these passages. Alternatively, NAGB might elect to enforce limits on the total length of all passages in a given test form, allowing somewhat greater variation in the length of individual passages than is implied by current specifications. Strict adherence to current length limits would mean that considerably more passage selection and item development would be needed in reading.

Updated Status, Including New Items

NAGB commissioned a group of scholars, designated as the Linkage Feasibility Team (LFT), to provide advice on how best to link scores on the VNT to the NAEP score scale and achievement level cutpoints (see discussion in Section 4). The LFT report, which was presented to NAGB at its May 1999 meeting, included a number of recommendations for changing the VNT test and item specifications to increase consistency with NAEP. For reading, the report recommended:

- increasing passage lengths;
- using text mapping procedures to ensure reading questions assess appropriate skills, not just surface level information;
- including more constructed response questions; and
- editing reading passages to eliminate "choppiness."

For mathematics, the recommendations included:

- increasing the number of constructed-response items to ensure that higher-order thinking skills are assessed;
- making the decision about calculator use and about use of gridded and drawn-response items; and
- redoing the content classifications of items.

Subsequently, AIR issued revised test specifications with updated counts of the number of items by content and format category to be included in each section of each test. The most significant change was that "gridded" items were eliminated from the mathematics tests because NAEP tryouts of this format type indicated that students had difficulties in filling out the grids appropriately. Gridded items developed for the VNT are being revised to be either 2- or 3-point constructed-response items, or distractors are being created to convert them into multiple-choice items. Other issues, most notably passage length limits, had not been fully resolved as this report is being completed, but further changes in the item and test specifications appear unlikely.

Mathematics

New information on the status of the mathematics items was received in July. The new file contained information on 202 items that were not included in the file received in April. Of these, 178 had "development year" set to 1999 and 24 had development year values of 1997 or 1998. One item from the April 1999 file had been dropped. In total, the number of active mathematics items had increased from 1,154 to 1,355.

The July file contained flags indicating which reviews had been completed, but it did not have information on the outcome of each review. In April, 217 items had been approved by NAGB "as is" and another 152 had been approved "with edits." Of the 217 not requiring further edits, 12 were scheduled for cognitive labs, and 26 had been flagged for edits in the scorability review, leaving 179 fully completed items (Table 3-1). The July file shows that 10 of the additional (pre-1999) items had been reviewed by NAGB, but the outcome of the review was not indicated. The April file also showed 5 items flagged as "drop" and 1 flagged as "revise and review again" in the NAGB review. These items are still on the current version of the file, but it is unclear whether they have been reviewed again

Of the 1,355 active mathematics items in the July file, 179 were fully complete and 1,176 items required further review. At the August 1999 NAGB meeting, the contractor indicated that 1,100 mathematics items would be reviewed by NAGB's appropriate subject-area committee between September and November of 1999. This plan suggests that virtually all of the 1,344 currently active items that had not been fully approved were expected to survive remaining AIR reviews and pass to NAGB for its final review. Table 3-4 shows the distribution of the "currently active" items by content strand and item format, compared with the number required for the pilot test. These results are subject to change depending on NAGB decisions regarding test specifications and on how the gridded items are rewritten and reclassified.

Reading

The number of reading passages has been increased from 95 to 108; see Table 3-5. However, there is still considerable lack of clarity over passage length requirements with many of the medium-length information passages flagged as either too short or too long. It is likely that NAGB will consider the length of passage pairs so that combining short and long passages may be acceptable. Also, six of the long literary passages were reclassified as information passages and as such are unusable under the current test specifications. Overall, there are 85 fully acceptable passages. This leaves a shortage of three passages in the medium literary category, but there are eight additional literary passages that are just a few words over the medium length limit.

The number of reading items has been increased from 1,392 to 1,848. The new items have not yet been extensively reviewed, so it is not possible to update the completion figures included in Table 3-2. Table 3-6 shows the number of active items by stance and item format. NAGB has reviewed all of the active passages and plans to review approximately 1,650 items between September and November 1999 in order to have 72 passages and a total of 1,104 appropriately distributed items for use in the pilot test.

In reviewing updated item information, the committee also noted that, as shown in Table 3-6, virtually all of the items designed to measure the "initial understanding" stance were multiple choice, while almost all of the items measuring the "reader/text" stance were constructed response. While this may be a logical approach, the committee has not seen a rationale for this differential use of item formats by reading stance and is not aware that this design has been specifically reviewed by reading content experts.

Findings and Recommendations

The item tracking system has been significantly improved since it was reviewed in the Phase I evaluation report (National Research Council, 1999b). Information on the new (1999) items and information on the results (or at least the occurrence) of various reviews for all items has been added to the database.

The committee is concerned, however, that the information in the database is not being used effectively by NAGB and its contractor. A key example of our concern is that the item development

TABLE 3-4 Mathematics Item Status (as of July 7, 1999)

Item Format[a] and Content Strand	Needed for Pilot	Active as of April	Active as of July	Additional Needed
ECR				
Algebra and Function	18	7	9	9
Geometry and Spatial	18	8	10	8
Other	None	20	21	0
Subtotal	36	35	40	17
SCR (3 points)				
Algebra and Function	18	26	41	0
Data, Statistics, Probability	18	25	33	0
Geometry and Spatial	18	26	33	0
Measurement	18	41	46	0
Number	36	43	44	0
Subtotal	108	161	197	0
SCR[b] (2 points)				
Algebra and Function	18	4	19	0
Data, Statistics, Probability	36	9	46	0
Geometry and Spatial	18	17	46	0
Measurement	18	11	39	0
Number	36	7	45	0
Subtotal	126	48	195[b]	0
GR[b]				
Algebra and Function	None	10	0	0
Data, Statistics, Probability	None	33	0	0
Geometry and Spatial	None	29	0	0
Measurement	None	26	0	0
Number	None	34	0	0
Subtotal	None	132	0	0
MC				
Algebra and Function	180	144	199	0
Data, Statistics, Probability	108	87	113	0
Geometry and Spatial	162	108	119	43
Measurement	162	157	185	0
Number	198	282	307	0
Subtotal	810	778	923	43
Total	1,080	1,154	1,355	60

[a]ECR = extended constructed response; SCR = short constructed response; GR = gridded; MC = multiple choice
[b]All of the gridded items were combined with the 2-point SCR items. Some of these items may be converted to MC items; however, there would still be a shortage of at least 15 geometry and spatial items for MC and 2-point SCR combined.

TABLE 3-5 Count of VNT Reading Passages by Type and Length (July 1999)

Type and Length	Total Needed	Total Written	Satisfactory Item Sets	Questionable Item Sets
Short Literary	12	13	13	0
Medium Literary	12	17	9	8[a]
Long Literary	12	13	13	0
Short Information	12	25	20	5[a]
Medium Information Pairs (1st of 2)	12	17	15	2[b]
Second Medium Information Pairs (2nd of 2)	12	17	15	2
Long Information	0	6	0	6[c]
Total Passages	72	108	85	23

[a]Word count exceeds the limit
[b]Too few extended constructed response or multiple choice items
[c]Items previously classified as "long literary" passages

TABLE 3-6 Reading Items by Stance and Format (as of July 1999)

Stance	Format[a] MC	SCR (2 points)	SCR (3 points)	ECR	Total	Needed for Pilot[b]	Ratio[c]
Initial Understanding	221	0	0	0	221	132.5	1.67
Developing and Interpretation	789	116	52	45	1,002	585.1	1.71
Reader/Text Interaction	11	122	32	34	199	110.4	1.80
Critical Stance	289	114	17	6	426	276.0	1.54
Total	1,310	352	101	85	1,848	1,104	1.67
Needed for Pilot Test[d]	876	120	60	48	1,104		
Ratio[c]	1.50	2.93	1.68	1.77	1.67		

[a]ECR = extended constructed response; SCR = short constructed response; MC = multiple choice.
[b]Distribution by stance is specified in the framework as initial understanding 12%; developing and interpretation 53%; reader/text interaction 10%; critical stance 25%.
[c]Ratio = total/needed for pilot
[d]Distribution by format is based on revised table of specifications, distributed at August 1999 NAGB meeting

subcontractors were given specifications for additional items without reference to item bank information on shortages in specific content and format categories. As a consequence, it appears that the contractor will still be a few items short of goals for the pilot test in one or two of the mathematics item categories. For reading, the contractor has not been able to (or not asked to) produce status counts that reflect the ties between items and passages. For each passage, NAGB will need to know where all of the associated items are in the review process. Currently, there is no field in the database for passages that shows whether one or both of two distinct item sets have passed each review stage.

RECOMMENDATION 3.1 NAGB should require regular item development status reports that indicate the number of items at each stage in the review process by content and

format categories. For reading, NAGB should also require counts at the passage level that indicate the status of passage reviews and the completeness of all of the associated items.

There are a large number of items scheduled for content, readability, achievement level, sensitivity, bias, and final NAGB review between August and November 1999. For each test, the contractor has developed more than the minimum number of required items in the event that some items do not survive all of these reviews. For the mathematics test, 1,080 of the current 1,344 items need to survive; for the reading test, 72 of the 126 current passages need to survive with two distinct item sets for use in the pilot. Plans are in place to complete each of the required review steps. In our interim report (National Research Council, 1999c), we recommended that the review process be accelerated to allow more time for AIR to respond to the reviews, and NAGB is now prepared to start its final review sooner than previously planned (September rather than November).

There is a sufficient overage of items for each test so that, assuming that the reviews are completed as scheduled, it should be possible to assemble 18 distinct forms of the mathematics test and 24 distinct forms of the reading test from the items surviving these reviews. Given that the number of mathematics items in some categories is already less than 18 times the number specified for each form, it is unlikely that each of the pilot test forms will exactly match the specifications for operational VNT forms, unless some items are included in multiple pilot test forms. In Chapter 4, we raise a question of whether additional extended constructed-response items should be included in the pilot test. Small shortages in the number of pilot test items in some item content and format categories might be tolerated or even planned for in order to accommodate potentially greater rates of item problems in other categories. However, the contractor has no basis for estimating differential rates at which items of different types will be dropped on the basis of pilot test result.

RECOMMENDATION 3.2 The rates at which each of the different item types survives each stage from initial content reviews through analyses of pilot test data should be computed. This information should be used in setting targets for future item development.

The contractor expects that, because of cognitive laboratory review, the survival rate for extended constructed-response items will be similar to that for other item types. Information from the current reviews and from the pilot test about the survival rates for different item types will provide both VNT and other test developers a better basis for estimating item survival rates in the future.

ITEM QUALITY

Assessing the quality of the VNT items was central to the committee's charge. The committee conducted a thorough study of the quality of VNT items that had been completed, or were nearly completed, at the time of our April 1999 workshop. Our review involved sampling available items, identifying additional content experts to participate in reviewing the items, developing rating procedures, conducting the item rating workshop, and analyzing the resulting data. A brief description of each of these steps is presented here, followed by the committee's findings and recommendations. More complete details of our item quality study can be found in Hoffman and Thacker (1999).

Evaluation Process

Sampling Completed Items

Using the item status information available for April 1999, we selected items to review, seeking to identify a sample that closely represented the content and format requirements for an operational test form. To assure coverage of the item domains, we sampled twice as many items as required for a form. Our sample thus included 120 mathematics items and 12 reading passages with 90 reading items, plus a small number of additional items to be used for rater practice sessions. Within each content and item format category, we sampled first from items that had already been approved "as is" by the NAGB review; in some cases, we had to sample additional items that had not yet been reviewed by NAGB but had been through the other review steps. We concentrated on items that had been included in the 1998 achievement-level matching exercise, did not have further edits suggested by the scorability review, and were not scheduled for inclusion in the 1999 cognitive laboratories. For reading, we first identified passages that had at least one completed item set. For medium-length informational passages, we had to select passage pairs together with intertextual item sets that were all relatively complete.

Table 3-7 shows the numbers of selected mathematics and reading items by completion status. Given the two-stage nature of the reading sample (item sets sampled within passage), we ended up with a smaller number of completed reading items than mathematics items. In our analyses, we also examined item quality ratings by level of completeness. (Additional details on the procedures used to select items for review can be found in Hoffman and Thacker [1999].) The items selected for review are a large and representative sample of VNT items that were then ready or nearly ready for pilot testing, but they do not represent the balance of the current VNT items, which are still under development.

Expert Panel

Our overall conclusions about item quality are based primarily on ratings provided by panels of five mathematics experts and six reading experts with a variety of backgrounds and perspectives, including classroom teachers, test developers, and disciplinary experts from academic institutions:

TABLE 3-7 Items for Quality Evaluation by Completion Status

	Current Item Status (Completeness)			
Subject	Approved by NAGB	Awaiting NAGB Review	Awaiting Edits or Cognitive Labs	Total Items Sampled
Mathematics	100	17	3	120
Reading	31	50	9	90

Mathematics

Pamela Beck	Test Developer; New Standards Mathematics Reference Exam, University of California, Oakland
Jeffrey Choppin	Teacher; Benjamin Banneker Academic High School, Washington, DC
Thomas Cooney	Committee Member and Professor of Mathematics, University of Georgia, Athens
Anna Graeber	Disciplinary Expert; Department of Curriculum and Instruction, University of Maryland, College Park
Catherine Yohe	Teacher; Williamsburg Middle School, Arlington, Virginia

Reading

Gretchen Glick	Test Developer; Defense Manpower Data Center, Seaside, California
John Guthrie	Committee Member and Professor, Department of Human Development, University of Maryland, College Park
Marjorie Lipson	Committee Member and Professor, Department of Education, University of Vermont, Burlington
Rosemarie Montgomery	Teacher/Disciplinary Expert; Retired English Teacher, Pennsylvania
Gale Sinatra	Disciplinary Expert; Department of Educational Studies, University of Utah, Salt Lake City
John Tanner	Test Developer; Assessment and Accountability, Delaware Department of Education, Dover

We allocated a total of 6 hours to the rating process, including initial training and post-rating discussion. Based on experience with the 1998 item quality ratings, we judged that this time period would be sufficient for each expert to rate the number of items targeted for a single VNT form, 60 math items or 45 reading items with associated passages.

Comparison Sample of NAEP Items

In addition to the sampled VNT items, we identified a supplemental sample of released NAEP 4th-grade reading and 8th-grade mathematics items for inclusion in the rating process, for two reasons. First, content experts will nearly always have suggestions for ways items might be improved. A set of items would have to be truly exemplary for a diverse panel of experts to have no suggestions for further improvement. Use of released and final NAEP items provides a reasonable baseline against which to compare the number of changes suggested for the VNT items. Second, NAGB has been clear and consistent in its desire to make the VNT as much like NAEP as possible; NAEP items thus provide a very logical comparison sample, much more appropriate than items from other testing programs. We also note that NAEP items provide the basis for a fairly stringent comparison because they have been administered to large samples of students, in contrast to the pre-pilot VNT items. In all, we sampled 26 NAEP math items and 3 NAEP reading passages with a total of 30 reading items.

We used released NAEP items, but we masked the identity of all items so that raters would not know which items were NAEP and which were VNT. Several of our raters were sufficiently familiar with NAEP that it may not have been possible for them to be fully blind to item source, but, we did make every possible effort to remove clues to each item's source.

Rating Booklet Design

In assigning items to rater booklets, we tried to balance the desire to review as many items as possible with the need to provide raters with adequate time for the review process and to obtain estimates of rater consistency levels. We assigned items to one of three sets: (a) those rated by all raters (common items), (b) those rated by two raters (paired items), and (c) those rated by only one rater (single items). Booklets were created (a different one for each rater) so as to balance common, paired, and single items across the books. Common item sets were incorporated into the review process in order to obtain measures of rater agreement and to identify outliers, those who consistently rated higher or lower than others.

For mathematics, each booklet contained three sets of common VNT items, targeted for three time slots: the beginning of the morning session (five items), the end of the morning session (ten items), and the end of the afternoon session (five items). For reading, the need to present items within passages constrained the common set of items to two VNT passages. These were targeted for presentation at the beginning (6 items) and end (11 items) of the morning rating sessions. The remaining VNT and NAEP items were assigned to either one or two raters. We obtained two independent ratings on as many items as possible, given the time constraints, in order to provide further basis for assessing rater consistency. The use of multiple raters also provided a more reliable assessment of each item, although our primary concern was with statistical inferences about the whole pool of items and not about any individual items. The items assigned to each rater were balanced insofar as possible with respect to content and format categories. (Further details of the booklet design can be found in Hoffman and Thacker [1999].)

Rating Task

The rating process began with general discussion among both rating panels and committee members to clarify the rating task. There were two parts of the rating task. First, raters were asked to provide a holistic rating of the extent to which the item provided good information about the skill or knowledge it was intended to measure. The panels started with a five-point scale, with each level tied to a policy decision about the item, roughly as follows:

1. flawed and should be discarded;
2. needs major revision;
3. acceptable with only minor edits or revisions;
4. fully acceptable as is; or
5. exceptional as an indicator of the intended skill or knowledge.

The panel of raters talked, first in a joint session, and later in separate sessions by discipline, about the reasons that items might be problematic or exemplary. Two kinds of issues emerged during these discussions. The first concerned whether the content of the item matched the content frameworks. For the mathematics items, the panel agreed that when the item appeared inappropriate for the targeted content strand, it would be given a code no higher than 3. For reading, questions about the target ability would be flagged in the comment field but would not necessarily constrain the ratings.

The second type of issue was described as craftsmanship. Craftsmanship concerns whether the item stem and response alternatives are well designed to distinguish between students who have or do not have the knowledge and skill the item was intended to measure. Items with obviously inappropriate incorrect choices are examples of poor craftsmanship.

The second part of the rating task involved providing comments to document specific concerns about item quality or specific reasons that an item might be exemplary. Major comment categories were identified in the initial panel discussion, and specific codes were assigned to each category to facilitate and standardize comment coding by the expert panelists.

After working through a set of practice items, each panel discussed differences in the holistic ratings or in the comment categories assigned to each item. Clarifications to the rating scale categories and to the comment codes were documented on flip-chart pages and taped to the wall for reference during the operational ratings. Table 3-8 lists the primary comment codes used by the panelists and provides a count of the frequency with which each of the codes was used by each of the two panels.

TABLE 3-8 Comment Coding for Item Rating

		Frequency of Use[a]	
Code Issue	Explanation	Mathematics	Reading
Content			
AMM	Ability mismatch (refers to mathematics content ability classifications)	17	0
CA	Content category is ambiguous: strand or stance uncertain	4	4
CAA	Content inappropriate for target age group	2	2
CE	Efficient question for content: questions gives breadth within strand or stance	3	0
CMM	Content mismatch: strand or stance misidentified	19	24
CMTO	More than one content category measured	8	2
CR	Rich/rigorous content	4	13
CRE	Context reasonable	0	3
CSL[b]	Content strand depends on score level	0	0
S	Significance of the problem (versus trivial)	12	1
Craftsmanship			
ART	Graphic gives away answer	0	1
B	Bias; e.g., gender, race, etc.	5	0
BD	Back-door solution possible: question can be answered without working the problem through	16	0
DQ	Distractor quality	32	55
II	Item interdependence	0	1
MISC	Miscellaneous, multiple	1	1
RR	Rubric, likelihood of answer categories: score levels do not seem realistically matched to expected student performance	6	4
STEM	Wording in stem	0	16
TD	Text dependency: question and text are too closely or loosely associated	3	13
TL	Too literal (correct answer matches a text sentence)	0	17
TQ	Text quality	14	1
VOC	Vocabulary: difficulty	0	3

[a]Used for VNT items only.
[b]Used only on two NAEP items.

Comment codes were encouraged for highly rated items as well as poorly rated items; however, the predominant usage was for items rated below acceptable. (See Hoffman and Thacker [1999] for a more complete discussion of the comment codes.)

Item Quality Rating Results

Agreement Among Panelists

In general, agreement among panelists was high. Although two panelists rating the same item gave the same rating only 40 percent of the time, they were within one scale point of each other approximately 85 percent of the time. In many of the remaining 15 percent of the pairs of ratings where panelists disagreed by more than one scale point, quality rating differences stemmed from different interpretations of test content boundaries rather than from specific item problems. In other cases, one rater gave the item a low rating, apparently having detected a particular flaw that was missed by the other rater.

Overall Evaluation

The results were generally positive: 59 percent of the mathematics items and 46 percent of the reading items were judged to be fully acceptable as is. Another 30 percent of the math items and 44 percent of the reading items were judged to require only minor edits. Only 11 percent of the math items and 10 percent of the reading items were judged to have significant problems.

There were no significant differences in the average ratings for VNT and NAEP items. Table 3-9 shows mean quality ratings for VNT and for NAEP reading and mathematics items and the percentages of items judged to have serious, minor, or no problems. Average ratings were 3.4 for VNT mathematics and 3.2 for VNT reading items, both slightly below the 3.5 boundary between minor edits and acceptable as is. For both reading and mathematics items, about 10 percent of the VNT items had average ratings that indicated serious problems. The proportion of NAEP items judged to have similarly serious problems was higher for mathematics (23 percent) and lower for reading (3 percent).

TABLE 3-9 Quality Ratings of Items

Subject and Test	Number of Items Rated[a]	Mean	S.D.	Percentage of Items with Scale Means of		
				Less Than 2.5[b]	2.5 to 3.5[c]	At Least 3.5[d]
Mathematics						
VNT	119	3.4	0.7	10.9	30.3	58.8
NAEP	25	3.1	0.9	23.1	30.8	46.2
Reading						
VNT	88	3.2	0.7	10.2	44.3	45.5
NAEP	30	3.2	0.5	3.3	50.0	46.7

[a]Two VNT reading items, one VNT mathematics item, and one NAEP mathematics item were excluded due to incomplete ratings.
[b]Items that at least need major revisions to be acceptable.
[c]Items that need minor revisions to be acceptable.
[d]Items that are acceptable.

The relatively high number of NAEP items flagged by reviewers as needing further work, particularly in mathematics, suggests that the panelists had high standards for item quality. Such standards are particularly important for a test such as the VNT. In NAEP, a large number of items are included in the overall assessment through matrix sampling. In the past, items have not been subjected to large-scale tryouts prior to inclusion in an operational assessment, and it is not uncommon for problems to be discovered after operational use so that the item is excluded from scoring. By contrast, a relatively small number of items will be included in each VNT form, and scores for individuals will be based on those few items, so the standards must be high for each one.

Evaluation of Different Types of Items

There were few overall differences in item quality ratings for different types of items, that is, by item format or item strand or stance. For the reading items, however, there was a statistically significant difference between items that had been reviewed and approved by NAGB and those that were still under review, with the items reviewed by NAGB receiving higher ratings. Table 3-10 shows comparisons of mean ratings by completeness category for both mathematics and reading items.

Specific Comments

The expert raters used specific comment codes to indicate the nature of the minor or major edits that were needed for items rated as less than fully ready (see Hoffman and Thacker, 1999). For both reading and math items, the most frequent comment overall, particularly for items judged to require minor edits, was "distractor quality" for both NAEP and VNT items. In discussing their ratings, the panelists were clear that this code was used when one or possibly more of the incorrect (distractor) options on a multiple-choice item was highly implausible and likely to be easily eliminated by respondents. This code was also used if two of the incorrect options were so similar that if one were correct, the other could not be incorrect. Other distractor quality problems included nonparallel options or other features that might make it possible to eliminate one or more options without really understanding the underlying concept.

TABLE 3-10 VNT Item Quality Means by Completeness Category

Subject and Test	Number of Items Rated	Mean[a]	S.D.	Percentage of Items with Scale Means of: Less Than 2.5[b]	2.5 to 3.5[c]	At Least 3.5[d]
Mathematics						
Review completed	99	3.4	0.8	12.1	31.3	56.6
Review in progress	20	3.5	0.5	5.0	25.0	70.0
Reading						
Review completed	31	3.4	0.6	3.2	41.9	54.8
Review in progress	57	3.1	0.7	14.0	45.6	40.3

[a]Reading means are significantly different at $p < .05$.
[b]Items that need major revisions to be acceptable.
[c]Items that need minor revisions to be acceptable.
[d]Items that are acceptable.

For both reading and mathematics items, the second most frequent comment code was "content mismatch." In mathematics, this code might indicate an item classified as an algebra or measurement item that seemed to be primarily a measure of number skills. In reading, this code was likely to be used for items classified as critical stance or developing an interpretation that were relatively literal or that seemed more an assessment of initial understanding. Reading items that were highly literal were judged to assess the ability to match text string patterns rather than gauging the student's understanding of the text. As such, they were not judged to be appropriate indicators of reading ability. In both cases, the most common problem was with items that appeared to be relatively basic although assigned to a more advanced content area.

For mathematics items, another frequent comment code was "backdoor solution," meaning that it might be possible to get the right answer without really understanding the content that the item is intended to measure. An example is a rate problem that is intended to assess students' ability to convert verbal descriptions to algebraic equations. For example, suppose two objects are travelling in the same direction at different rates of speed, with the faster object following the slower one, and the difference in speeds is 20 miles per hour, and the initial difference in distance is also 20 miles. Students could get to the answer that it would take 1 hour for the faster object to overtake the slower one without ever having to create either an algebraic or graphical representation of the problem. The expert mathematics panelists also coded a number of items as having ambiguous ability classifications. Items coded as problem solving seemed sometimes to assess conceptual understanding, while other items coded as tapping conceptual understanding might really represent application. By agreement, the panelists did not view this as a significant problem for the pilot test, so many of the items flagged for ability classifications were rated as fully acceptable.

For reading items, the next most frequent code was "too literal," meaning that the item did not really test whether the student understood the material, only whether he or she could find a specific text string within the passage.

Conclusions and Recommendations

With the data from item quality rating panels and other information provided to the committee by NAGB and AIR, the committee reached a number of conclusions about current item quality and about the item development and review process. We stress that there are still no empirical data on the performance and quality of the items when they are taken by students, and so the committee's evaluation is necessarily preliminary.

Most testing programs collect empirical (pilot test) item data at an earlier stage of item development than has been the case with the VNT. The key test of whether items measure intended domains will come with the administration of pilot test items to large samples of students. Data from the pilot test will show the relative difficulty of each item and the extent to which item scores provide a good indication of the target constructs as measured by other items. These data will provide a more solid basis for assessing the reliability and validity of tests constructed from the VNT item pool.

We conclude that the quality of the completed items is as good as a comparison sample of released NAEP items. Item quality is significantly improved in comparison with the items reviewed in preliminary stages of development a year ago.

As described above, the committee and other experts reviewed a sample of items that were ready or nearly ready for pilot testing. Average quality ratings for these items were near the boundary between "needs minor edits" and "use as is" and were as high as or higher than ratings of samples of released NAEP items. The need for minor edits does not affect the readiness of items for pilot testing.

Although quality ratings were high, the expert panelists did have a number of suggestions for improving many of the items. One frequent concern was with the quality of the distractors (incorrect options) for multiple-choice items. While distractor problems were coded as a minor editorial problem, such problems can seriously degrade the quality of information obtained during pilot testing. One example that typifies the kinds of problems that stem from poor distractor quality would be not including "2a" as an option for a question asking the value of "a times a." Clearly, such an omission would affect item difficulty estimates and might lead to a conclusion that the item was easy, when in fact it was not.

The match to particular content or skill categories was also a frequent concern. More serious flaws included reading items that were too literal or mathematics items that did not reflect significant mathematics, possibly because they had back-door solutions. However, the rate for flagged items was not higher than the rate at which released NAEP items were similarly flagged. Many of the minor problems, particularly distractor quality issues, are also likely to be found in the analysis of pilot test data.

RECOMMENDATION 3.3 Item quality concerns identified by reviewers, such as distractor quality and other "minor edits," should be carefully addressed and resolved by NAGB and its contractor prior to inclusion of any items in pilot testing.

In the best of circumstances, items to be pilot tested should be as perfected as possible so that the student response data will lead to minimal changes. The uncertainty surrounding the VNT and the rapid development schedules provide very little time for further testing of edited items or for evaluating the effects of changes in items on the test forms as a whole. It is reasonable to assume that the more perfected the piloted items are, the higher the item survival rate will be. It will also be easier to assemble all operational VNT test forms to meet the same statistical specifications if items are not revised following pilot testing.

MATCHING VNT ITEMS TO NAEP ACHIEVEMENT-LEVEL DESCRIPTIONS

In the interim Phase I evaluation report (National Research Council, 1998a:6), the NRC recommended "that NAGB and its contractors consider efforts now to match candidate VNT items to the NAEP achievement-level descriptions to ensure adequate accuracy in reporting VNT results on the NAEP achievement-level scale." This recommendation was included in the interim report because it was viewed as desirable to consider this matching before final selection of items for inclusion in the pilot test. The recommendation was repeated in the final Phase I report (National Research Council, 1999b:34): "NAGB and the development contractor should monitor summary information on available items by content and format categories and by match to NAEP achievement-level descriptions to assure the availability of sufficient quantities of items in each category."

Although the initial recommendation was linked to concerns about accuracy at different score levels, the Phase I report was also concerned about the content validity of achievement-level reporting for the VNT. All operational VNT items would be released after each administration, and if some items appeared to measure knowledge and skills not covered by the achievement-level descriptions, the credibility of the test would suffer. There will also be a credibility problem if some areas of knowledge and skill in the achievement-level descriptions are not measured by any items in a particular VNT form, but this problem is more difficult to address in advance of selecting items for a particular form. Finally, there also would be validity questions if a student classified at one achievement level answered correctly most questions matched to a higher level or missed most questions that matched the achievement description for a lower level.

Contractor Workshop

In fall 1998 the test development contractor assembled a panel of experts to match then-existing VNT items to the NAEP achievement levels. Committee and NRC staff members observed these ratings, and the results were reported at the committee's February workshop (American Institutes for Research, 1999b). The contractor's main goal in matching VNT items to NAEP achievement levels was to have an adequate distribution of item difficulties to ensure measurement accuracy at key scale points. The general issue was whether item difficulties matched the achievement-level cutpoints. However, there was no attempt to address directly the question of whether the content of the items was clearly related to the "descriptions" of the achievement levels. The expert panelists were asked which achievement level the item matched, including a "below basic" level for which there is no description; they were not given an option of saying that the item did not match the description of any of the levels.

In matching VNT items to achievement levels, the treatment of constructed-response items with multiple score points was not clarified. The score points do not correspond directly to achievement levels, since scoring rubrics are developed and implemented well before the achievement-level descriptions are final and the cutpoints are set. Nonetheless, it is possible, for example, that "basic" or "proficient" performance is required to achieve a partial score, while "advanced" performance is required to achieve the top score for a constructed-response item. Committee members who observed the process believed that multipoint items were generally rated according to the knowledge and skill required to achieve the top score.

The results of the contractor's achievement-level matching varied by subject. For reading, there was reasonably good agreement among judges, with two of the three or four judges agreeing on a particular level for 94 percent of the items. Only 4 of the 1,476 reading items for which there was agreement were matched to the "below basic" level. About half of the items matched the proficient level, a quarter of the items were matched to the basic level, and a quarter to the advanced level. Based on these results, the contractor reports targeting the below basic level as an area of emphasis in developing further reading items.

For mathematics, there was much less agreement among the judges: the three or four panelists each selected a different achievement level (of the four possible). In addition, roughly 10 percent of the mathematics items were matched to the "below basic" level, for which there was no written description.

In an effort to begin to address the content validity concerns about congruence of item content and the achievement-level descriptions, we had our reading and math experts conduct an additional item rating exercise. After the item quality ratings, they matched the content of a sample of VNT items to descriptions of the skills and knowledge required for basic, proficient, or advanced performance. The descriptions used in this exercise were a tabular arrangement of the words in the descriptions approved by NAGB. Appendix B shows the achievement-level descriptions for 4th-grade reading and for 8th-grade mathematics that have been approved by NAGB and the reorganization of these descriptions used by the panelists. Panelists were asked whether the item content matched *any* of the achievement level descriptions and, if so, *which ones*. Thus, for multipoint items it was possible to say that a single item tapped basic, proficient, and advanced skills.

In general, although the panelists were able to see relationships between the content of the items and the achievement-level description, they had difficulties in making definitive matches. In mathematics, the few items judged not to match any of the achievement-level descriptions were items that the panelists had rated as flawed because they were too literal or did not assess significant mathematics.

The panelists expressed concern about the achievement-level descriptions to which the VNT items were matched. The current descriptions appear to imply a hierarchy among the content areas that the panelists did not endorse. In reading, for example, only the advanced achievement-level description talked about critical evaluation of text, which might imply that all critical stance items were at the advanced level. A similar interpretation of the descriptions could lead one to believe that initial interpretation items should mostly be at the basic level. The panelists pointed out, however, that by varying passage complexity and the subtlety of distinctions among response options, it is quite possible to construct very difficult initial interpretation items or relatively easy critical stance items. They noted, for example, an item that used a very simple literary device (capitalization of all letters of one word); the item would have to be classified as advanced, because literary devices are limited to the advanced achievement level. Raters were dismayed at the prospect of categorizing such a simplistic item as advanced. Perhaps a better approach for the VNT would be to develop descriptions of basic, proficient, and advanced performance for each of the reading stances and to provide a description of complexity and the fineness of distinctions that students would be expected to handle at each level. This approach would provide more useful information to parents and teachers about students' skills.

For mathematics, there were similar questions about whether mastery of concepts described under advanced performance necessarily implied that students also could perform adequately all of the skills described as basic. For these, too, the panelists suggested, it would be useful, at least for informing instruction, to describe more specific expectations within each of the content strands rather than relying on relatively "content-free" descriptions of problem-solving skills.

The committee was concerned about the completeness with which all areas of the content and achievement-level descriptions are covered by items in the VNT item pool. Given the relatively modest number of completed items, it is not possible to answer this question at this time. In any event, the primary concern is with the completeness of coverage of items in a given test form, not with the pool as a whole. The current content specifications will ensure coverage at the broadest level, but assessment of completeness of coverage at more detailed levels must await more complete test specifications or the assembly of actual forms.

The committee did not attempt to address the issue of the validity of the achievement-level descriptions as they are used in NAEP. A number of prior reviews have questioned the process used to develop the NAEP achievement levels—both the scale points that operationally divide one level from the next and the description of the knowledge and skills associated with performance at the basic, proficient, and advanced levels (National Research Council, 1999a; National Academy of Education, 1993). Other experts have defended the process used in developing NAEP achievement levels (see, e.g., Hambleton et al., 1999). The issue of whether the standards are too high or too low is a matter of NAGB policy and not something the committee considered within its charge. Rather, the committee focused on whether the content of the VNT items appeared to match the descriptions developed by NAGB for reporting results by achievement levels.

Conclusions and Recommendations

In reviewing efforts by NAGB and its contractor to match VNT items to NAEP achievement-level descriptions, the committee's overall conclusion is that these efforts have been helpful in ensuring a reasonable distribution of item *difficulty* for the pilot test item pool, but they have not yet to begun to meet the need to ensure a match of item *content* to the descriptions of performance at each achievement level.

As described above, the achievement-level matching conducted by the development contractor focused on item difficulty and did not allow the raters to identify items that did not match the content of any of the achievement-level descriptions. Also, for mathematics, there was considerable disagreement among the contractor's raters about the achievement levels to which items were matched.

The committee's own efforts to match item content to the achievement-level descriptions led to more concern with the achievement-level descriptions than with item content. The current descriptions do not provide a clear picture of performance expectations within each reading stance or mathematics content strand. The descriptions also imply a hierarchy among skills that does not appear reasonable to the committee.

The match between item content and the achievement-level descriptions and the clarity of the descriptions themselves will be particularly critical to the VNT. Current plans call for releasing all of the items in each form, immediately after their use. Unlike NAEP, individual VNT scores will be given to students, parents, and teachers, which will lead to scrutiny of the results to see how a higher score might have been obtained. The achievement-level descriptions will have greater immediacy for teachers seeking to focus instruction on the knowledge and skills outlined as essential for proficiency in reading at grade 4 or mathematics at grade 8. Both the personalization of the results and the availability of the test items suggest very high levels of scrutiny and the consequent need to ensure that the achievement-level descriptions are clear and that the individual items are closely tied to them.

RECOMMENDATION 3.4 The contractor should continue to refine the achievement-level matching process to include the alignment of item *content* to achievement-level descriptions, as well as the alignment of item *difficulty* to the achievement-level cutpoints.

RECOMMENDATION 3.5 The achievement-level descriptions should be reviewed for usefulness in describing specific knowledge and skill expectations to teachers, parents, and others with responsibility for interpreting test scores and promoting student achievement.

The most justifiable scientific model of reading at grade 4 consists of a set of lower-level and higher-level processes operating together. Basic reading comprehension requires both higher and lower processes (Kintsch, 1998). The processes are interactive. Processes such as word recognition, recalling word meanings, and understanding sentences are necessary prerequisites for comprehension and construction of knowledge from text (Lorch and van den Broek, 1997). In addition, higher-level processes of using background knowledge, making inferences, and evaluating new information are central to comprehension (Graesser, Singer, and Trabasso, 1994). Furthermore, these higher-level processes can also increase lower-level processes. Higher and lower processes influence each other in top-down and bottom-up mechanisms (Anderson and Pearson, 1984). Therefore, tests should represent the higher-level processes of using knowledge, making inferences, and judging critically at all levels. These higher-level processes should be present at the basic achievement level as well as the proficient and advanced levels of the VNT and NAEP.

It is not justified to state that students at the basic level of NAEP have sentence comprehension or initial understanding, but not critical evaluation in reading. Rather, students at the basic level have relatively less developed competencies in all processes, including word recognition, making inferences, knowledge use, and critical evaluation, which can be applied to relatively simple texts. Students at the advanced level have acquired these same reading competencies to an expert level. They possess more complex forms of these competencies, and they can use them to comprehend more complex texts. The descriptions of achievement levels should reflect the widely accepted interactive model of reading.

DOMAIN COVERAGE

A very key question about the quality of the VNT items, in addition to their individual fit to the test frameworks, is whether in the aggregate they cover completely the intended frameworks. Given the committee's concern about unintended implications about content categories and proficiency levels, we decided to assess whether there is good coverage of each content, process, and proficiency category. Are there relatively advanced items on developing initial interpretations in reading or computation in mathematics? Are there more basic items on developing a critical stance (reading) or in probability and statistics (mathematics)? Unfortunately, these are questions that the committee cannot answer at this time due to the time constraints on our work. New items, written to fill in perceived gaps in the domain coverage, had not yet been reviewed, and a relatively high proportion of the original items were also not fully completed.

Importance of Coverage

The question of domain coverage that concerns the committee is not just a matter of whether the item bank, as a whole, covers all of the intended content, process, and proficiency categories. The key question is whether each and every test form includes a reasonable sampling of items from each of these categories. This is an important question because the planned release of all of the test items after operational use will communicate the intended domain to teachers, parents, curriculum developers, and others much more concretely than the more general descriptions included in the test frameworks and test and item specifications.

At its final meeting, the committee reviewed a document from AIR entitled "Technical Specifications, Revisions as of June 18, 1999." This document outlines criteria and procedures for selecting items to be administered and for assembling forms from these items. In this document, the contractor specifies the acceptable ranges for p-values (item difficulty estimates) and biserial correlations (for item scores with total test scores and for distractors with the total test score). The test blueprint is also specified for reading and math. The contractor notes: "After all forms are assembled, the final evaluation is conducted for all forms at the form level to determine whether all the forms are parallel and meet the form assembly criteria" (American Institutes for Research, 1999i:11).

The committee stresses that such an evaluation of forms is an essential part of the process and should be given a substantial amount of time, expertise, and resources. It may be advisable to have an expert panel with content and psychometric members as well as teachers evaluate the forms for both the reading and the mathematics tests. Content panels involved in item revision and form construction should include psychometricians, curricular specialists, and teachers. For mathematics items, the panel should also include mathematics educators and college or university mathematics faculty. For reading items, reading educators and reading researchers should be included. The cognitive labs might be considered as sources for review and revision of forms. Multiple forms should be examined simultaneously to ensure that the content frameworks and achievement levels are comparably represented.

The stated purpose (National Assessment Governing Board, 1999e:5) of the VNT is "to measure individual student achievement in 4th grade reading and 8th grade mathematics, based on the content and rigorous performance standards of the National Assessment of Educational Progress (NAEP), as set by the National Assessment Governing Board (NAGB)." The intended use (p. 9) is "to provide information to parents, students, and authorized educators about the achievement of the individual student in relation to the content and the rigorous standards for the National Assessment, as set by the National Assessment Governing Board for 4th grade reading and 8th grade mathematics."

There is reason to be concerned that the VNT, in its emerging development, may result in an assessment that is not challenging enough to meet the stated purpose and intended use of the test. This concern was expressed by NAGB's Linking Feasibility Team (Cizek et al., 1999:60): "Compared to the NAEP, the VNT-R [VNT reading test] appears to have a disproportionate number of questions that ask for trivial or insignificant information." Furthermore, "more constructed response questions should be added to the VNT-R. This will increase the number of higher order thinking items on the test" (Cizek et al., 1999:91).

Conclusions and Recommendations

The results of our item review pointed to similar areas of concern about domain coverage (see Hoffman and Thacker, 1999). Although the ratings of the VNT items and NAEP were generally similar, 14.5 percent of the panelists' comments were coded as "too literal," while none of the NAEP items were coded this way. The majority of items for the stances labeled "reader/text connection" and "critical stance" were rated as involving at least some difficulty (67% and 59%, respectively; see Hoffman and Thacker, 1999:Table 14). Similarly, in the qualitative reviews, reading panelists noted that many items were merely fact-finding from the passage and did not really match any of the stances (see Hoffman and Thacker, 1999:32). For reading, when items were problematic, the greatest frequency of comments were those having to do with "content rigor" (46.7% of those rated "2", and 33.3% of those rated "3") and "too literal" (26% of those rated "2" and 70.4% of those rated "3"). The other most frequently named concern was "distractor quality" (25.9% of those rated "2" and 70.4% of those rated "3").

For mathematics, panelists commented that it seemed like there were a lot of easy items with 4 ratings—the pool of completed items seems either easy or not significant mathematics (Hoffman and Thacker, 1999:31). If the VNT is to be a useful assessment, it must provide information not otherwise available, particularly in areas where there have been challenges to the rigor of the state standards.

RECOMMENDATION 3.6 Test blueprints should be expanded to indicate the expected number of items at each achievement level for each content area (reading stance or mathematics content strand) for each form of the test. Insofar as possible, items at each achievement level should be included for each content area.

4

Technical Issues in Test Development

The year 1 evaluation encompassed an expansive review of pilot test and field test design features, including plans for numbers of items developed, numbers of examinees responding to each item, student sampling procedures, equating the various forms, and analyzing the resulting data. This year 2 evaluation focuses on the following:

- the extent to which the design for pilot testing will result in items that represent the content and achievement-level specifications, are free of bias, and support test form assembly;
- plans for the implementation of VNT pilot testing;
- plans for assembling field test forms likely to yield valid achievement-level results; and
- technical adequacy of revised designs for field testing, equating, and linking.

The committee's interim report (National Research Council, 1999c) focused on the first three topics. Since that time, a report was issued by NAGB's Linkage Feasibility Team (LFT) on issues associated with linking VNT scores to the NAEP scale and the NAEP achievement-level cutpoints on that scale. The committee reviewed this report and discussed it with NAGB staff and one of the LFT report authors at its July 1999 meeting. Committee members also observed the discussions of AIR's VNT Technical Advisory Committee at its meeting in June.

This final report includes an expanded discussion of the first three topics as well as the committee's findings and recommendations on issues associated with linking VNT scores to the NAEP scale. The committee reviewed the following documents:

- Linking the Voluntary National Tests with NAEP and TIMSS: Design and Analysis Plans (American Institutes for Research, 1998g)
- Designs and Item Calibration Plan for the 1999 Pilot Test (American Institutes for Research, 1998f)

- Designs and Item Calibration Plans for Including NAEP Item Blocks in the 1999 Pilot Test of the VNT (American Institutes for Research, 1998e)
- Proposed Plan for Calculator Use (American Institutes for Research, 1998j)
- Field Test Plans for the VNT: Design and Equating Issues (American Institutes for Research, 1999c)
- Score Reporting, Draft (American Institutes for Research, 1998l)
- Test Utilization, Draft (American Institutes for Research, 1998n)
- Score Reporting, Scoring Examinees, and Technical Specifications: How Should These Be Influenced by the Purposes and Intended Uses of the VNT? (American Institutes for Research, 1999g)
- Selected Item Response Theory Scoring Options for Estimating Trait Values (American Institutes for Research, 1999h)
- Final Report of the Study Group Investigating the Feasibility of Linking Scores on the Proposed Voluntary National Tests and the National Assessment of Educational Progress (Cizek et al., 1999)
- Revised Plans for Linking the Voluntary National Test with NAEP (Johnson, 1999a)
- Using Social Moderation to Link the VNT to NAEP (Paulsen, J1999)
- VNT: Forms Assembly Procedures and Technical Specifications for the VNTs (AIR, for contract #RJ97153001, 1999l)
- An Evaluation of the VNT Pilot Test Design (Reckase, 1999)
- VNT Pilot Design Features (Johnson, 1999b)
- Evaluation of VNT Pilot Test Design (Hanson, 1999)
- Synthesis Paper on VNT Pilot Test Design Features (Ercikan, 1999)
- Technical Specifications, Revisions as of June 18, 1999 (American Institute for Research, 1999i)

Many of these documents describe design options on which the Governing Board has not yet taken a definitive position. This report provides comments and recommendations for NAGB's consideration.

PILOT TEST PLANS

Forms Design

Key features of the pilot test forms design are the use of school clusters, the use of hybrid forms, NAEP anchor forms, and item calibration procedures. Each of the first three features affects the item calibration plan.

Use of School Clusters

Current plans for pilot testing call for each participating school to be assigned to one of four school clusters. School clusters are used in the data collection design to minimize item exposure and to improve item security. Schools are assigned to clusters using a combination of random and systematic stratified sampling to maximize the equivalence of examinees across clusters. Forms are assigned to schools within each school cluster so that all schools in a given cluster are administered the same set of pilot forms. The forms are then distributed within each classroom in a given school in a systematic manner so that the examinees completing the different forms within each school cluster will be randomly equivalent.

The introduction of school clusters increases the complexity of the data collection design—from a random-groups design to a common-item nonequivalent-groups design. The larger the number of school clusters used, the fewer the number of items that will be threatened by a security breach. However, as the number of school clusters increases, creating equivalent school clusters becomes more difficult, and the analyses become more complex.

We conclude that the choice of four school clusters is a good compromise between the need to minimize item exposure and the need to produce accurate item parameters. It is probably the smallest number needed to minimize loss in the event of compromise at a particular school while also minimizing the complexity for administration and analysis. This conclusion is consistent with recommendations in the papers on pilot test design commissioned by AIR (Johnson, 1999b; Hanson 1999; Reckase, 1999).

Use of Hybrid Forms

The pilot test design calls for the creation of a number of "hybrid forms" that are comprised of the first half (45-minute session) of one form (e.g., 1a) paired with the second half of another form (e.g., 2b). Each pilot test form will resemble an operational "final" form insofar as possible with respect to length and administration time, distribution of items by content and format, and distribution of items with respect to other factors (such as calculator use). The use of hybrid or overlapping forms in the data collection design has merit because it permits accurate estimation of item parameters even if the groups within school clusters turn out not to be equivalent. Another advantage of the hybrid design is that it will allow intact NAEP blocks to be combined with VNT half-test blocks, which will provide a basis for comparing VNT and NAEP item difficulties and putting the VNT item parameters on the NAEP scale. To the extent that the NAEP blocks cover the content domain, it also will allow an assessment of the extent to which the VNT and NAEP measure the same areas of knowledge. Thus, we agree with the plan that a hybrid or other overlapping forms design be used in the pilot test.

Generally, data for item calibration must be collected either by administering different collections of items to equivalent samples of students or by administering overlapping collections of items to different samples of students. In the proposed pilot test design, parameters for items appearing in different forms must be placed on the same scale. Without the hybrid forms, the only way to do this is to assume that the random assignment of forms to students within school clusters has worked and has created equivalent groups of students taking each form. This assumption is somewhat risky because logistical problems commonly arise during test administration, leading to unintended deviations from the forms distribution plan. Such deviations affect the equivalence of the groups receiving each form. Thus, a more conservative procedure is to use an overlapping forms design, such as the one proposed by AIR, that provides for different groups of individuals within each school cluster to take overlapping forms.

The rationale for the use of the overlapping forms design is not clearly described in the materials we reviewed. The contractor needs to provide a better explanation for incorporating the hybrid forms design into the pilot study data collection design. The contractor suggests that two advantages of the hybrid forms design are that it provides some protection against violations of the assumption of local independence and that it permits investigation of item context effects. However, violations of local independence are most likely to occur within item sets, and item context effects are most likely to occur because of changes in item order within a test section. Thus, both of these effects are more likely to occur among items within a session than across sessions. It is therefore unclear to us how the proposed design will provide protection against these particular effects, although we endorse the use of hybrid forms, as noted above, for other reasons.

NAEP Anchor Forms

A major feature of the proposed VNT is that student performance will be reported in terms of NAEP achievement levels. To facilitate linking the two assessments, the most recent version of the pilot test design calls for the inclusion of NAEP item blocks in two of the four school clusters. The proposed item calibration plan calls for the estimation of NAEP item parameters along with VNT item parameters and, thus it implicitly assumes that NAEP and VNT measure the same content constructs. This assumption can be questioned since the distribution of item formats differs for the two assessments (e.g., differing numbers of constructed-response and multiple-choice items). Data for the groups of examinees in the proposed design who take VNT items in one session and NAEP items in the other session (e.g., 1a,Nb) can be used to assess the extent to which VNT and NAEP measure the same skills. For example, correlations between scores for two VNT sessions, between two NAEP sessions, and between a VNT session and a NAEP session can be computed and compared. We strongly support the inclusion of NAEP blocks in the pilot test design to provide data on the feasibility of a common calibration of VNT and NAEP items as a means of linking the two scales (see discussion of linkage issues below).

Item Calibration

Item calibration refers to the procedures used for estimating item parameters or characteristics of items, such as difficulty level. For NAEP (and proposed for VNT), item calibration is accomplished by using procedures based on item response theory (IRT), a statistical model that expresses the probability of getting an item correct as a function of the underlying ability being measured. Item characteristics are group dependent, that is, an item may appear easy or hard depending on the ability level of the group taking the item. Thus, to compare the difficulty parameter estimates of items that were administered to different sets of examinees, it is necessary to place (or link) the different sets of item parameter estimates on a common scale. For VNT items, the desire is to link the item parameters to the NAEP scale. The item calibration and linking process is technically complex, and the committee's findings and suggestions are described below in a technical manner in order to be of the most use to the test developers.

A variety of procedures can be used for obtaining item parameter estimates that are on the same scale using the common-item nonequivalent-groups design. The contractor presents three options: (1) simultaneous (one-stage) calibration, (2) two-stage calibration and linking, and (3) the Stocking-Lord test characteristic curve (TCC) transformations (American Institutes for Research, 1998e). The contractor states that a disadvantage of the Stocking-Lord procedure is that "the procedure may result in the accumulation of relatively large amounts of equating error, given the large number of 'links' in the chain of equatings required to adjust the test characteristic curves of some of the non-anchor forms. Also, it may be prohibitively time-consuming given the large number of computations required" (American Institutes for Research, 1998e:11). This rationale is also presented as the primary reason for making the Stocking-Lord procedure the least preferred method for putting item parameters on a common scale.

There are several alternative ways in which the Stocking-Lord TCC procedure can be implemented for the VNT pilot test design. Two options that deserve more consideration are presented below. Both use a computer program developed by the Educational Testing Service (ETS) called PARSCALE, the program that is used to produce item parameter estimates for NAEP.

In option 1, for each school cluster, perform one PARSCALE run that simultaneously calibrates all items within that cluster. Select a school cluster to use as the base scale (say, cluster 1). Use the item parameters for the common items (i.e., 1a, 1b, 2a, 2b) to compute a scale transformation from cluster 2 to cluster 1 and apply that transformation to the item parameter estimates for all items within cluster 2. Repeat this process for clusters 3 and 4. This option produces one set of item parameters for all items in the non-anchor forms, but it results in four sets of item parameters for the items in the anchor forms.

In option 2, perform one PARSCALE run for the anchor forms, combining data across school clusters, and then perform the four within-cluster PARSCALE runs described in option 1. The base scale is defined using the item parameter estimates for the anchor items from the across-cluster run. In each cluster, a scale transformation is computed using the item parameter estimates from the within-cluster run and the item parameter estimates from the across-cluster run.

Option 1 for implementing the Stocking-Lord TCC procedure requires three scale transformations, and option 2 requires four scale transformations. Neither option requires a "large" number of transformations, and both are as easy to implement as the two-stage calibration and linking procedure.

The across-cluster PARSCALE run in option 2 is the same as the first stage of the two-stage calibration and linking procedure proposed by the contractor. The four within-cluster PARSCALE runs in options 1 and 2 are similar to stage two of the two-stage calibration and linking procedure, with the exception that the item parameters for the anchor items are estimated rather than fixed. An advantage of the Stocking-Lord TCC procedure over the two-stage calibration and linking procedure is that the multiple sets of parameter estimates for the anchor forms can be used to provide a check on model fit. Consequently, we suggest that the contractor select the calibration procedure that is best suited to the final data collection design, is compatible with software limitations, and permits item-fit analyses. To further assess the degree to which VNT and NAEP measure similar areas of knowledge, calibrations for VNT items could be performed and compared both with and without the NAEP items.

RECOMMENDATION 4.1 Pilot test plans should include school clusters, overlapping (hybrid) forms design, and NAEP anchor forms, as currently planned. In addition, the contractor should select the calibration procedure that is best suited to the final data collection design and in accord with software limitations and should plan to conduct item-fit analyses.

Forms Assembly and Item Survival Rates

A key question addressed in the Phase I report was whether the number of items to be pilot tested is large enough to enable the assembly of six high-quality forms. The primary purpose of the pilot test is to use statistical information to identify "flawed" items. Flawed items are set aside for further revision or dropped altogether from further consideration. Items that are not flagged as flawed are said to "survive" the pilot test phase.

In mathematics, 18 forms of VNT items will be pilot tested in order to support assembly of 6 operational forms. The minimum survival rate for items in each mathematics content and format category is one-third. The average survival rate, however, must be quite a bit higher. The developers need a pool that is larger than six forms of items to draw from in order to select items for each form that meet difficulty distribution and information targets, as well as content and format requirements. On the basis of experience with other programs and because of the extensive editorial review of the VNT items prior to pilot testing, we believe it is reasonable to expect at least a two-thirds survival rate for the mathematics items, yielding two items for every one needed in the final forms.

For reading, the situation is a bit more complex. A total of 24 forms of reading items will be included in the pilot test, and each reading passage will be included twice with distinct sets of items. Overall, there will be 72 passages in the pilot test and 36 passages in the six operational forms constructed for the field test. Thus, the minimum survival rate for each purpose and length category of reading passage is 50 percent. For each type of passage, there must be a specified number of questions: five multiple-choice questions for short passages (literary or informational), and seven multiple-choice, two short constructed-response, and one extended constructed-response question for each long literary passage. Medium information passages have been developed in pairs: for each pair, there must be a set of three multiple-choice intertextual items asking questions that draw on material from both passages in the pair. Each passage or passage pair is being tried out with two independent sets of items. For the passage to be included in an operational form, at least 50 percent of the items, for each format type, must survive pilot test screening.

The contractor has indicated that pilot test forms will be assembled to match the target form specifications to the extent possible. Although the idea of creating pilot test forms that resemble operational forms is reasonable, it implies an equal survival rate for various types of items. This may not be the case, since constructed-response items have had lower survival rates than multiple-choice items in other programs and in NAEP. The materials we reviewed did not specify the expected survival rates for the various item types, nor did they discuss the rationale for determining item production or item survival rates. These issues were discussed at our July meeting with AIR staff, who responded that because the constructed-response items had gone through cognitive laboratory analysis, the survival rate for these items is expected to be as high or higher than that of the multiple-choice items. Because VNT item development is necessarily unique in many ways, neither the developers nor the committee have an empirical basis for estimating survival rates. We concur with the developer's judgment that the overall number of items to be pilot tested appears reasonable and repeat our hope that further edits of distractor quality will reduce the number of items dropped after pilot testing.

RECOMMENDATION 4.2 Information regarding expected item survival rates from pilot to field test should be stated explicitly, and NAGB should consider pilot testing additional constructed-response items, given the likelihood of greater rates of problems with these types of items than with multiple-choice items.

Pilot Test Analyses

The materials we reviewed at our July meeting included specifications for screening items according to item difficulty levels and item-total correlations based on the pilot test data (American Institutes for Research, 1999i). More recent documents have added plans for screening multiple-choice items for appropriate statistics on each incorrect (distractor) option, as well as for the correct option (American Institutes for Research, 1999l). Plans should be extended to include screening items on the basis of model fit for item response theory.

Differential Item Functioning

Additional specifications are needed for the ways in which items will be screened for differential item functioning (DIF) and the ways DIF results will be used in making decisions about the items.

Differential item functioning refers to the situation that occurs when examinees from different groups have differing probabilities of getting an item correct, after being matched on ability level. For DIF analyses, examinees may be placed into groups according to a variety of characteristics, but often gender and ethnicity are the groupings of interest. Groupings are created so that one is the *focal* group (e.g., Hispanics) and the other is the *referent* group (e.g., whites). The total test score is generally used as the ability measure on which to match individuals across groups.

DIF analyses compare the probabilities of getting an item correct for individuals in the focal group with individuals in the referent group who have the same test score. Items for which the probabilities differ significantly (or for which the probabilities differ by a specified amount) are generally "flagged" and examined by judges to evaluate the nature of the differences. Good test development practice calls for symmetric treatment of DIF-flagged items. That is, items are flagged whenever the probability of correct response is significantly lower for one group than for the other after controlling for total test score, whether the lower probability is for the focal group or for the referent group.

A number of statistical procedures are available for use in screening items for DIF. The contractor has proposed to use the Mantel-Hanszel method for the pilot test data and methods based on item response theory for the field test data. The sampling plan will allow for comparisons based on race/ethnicity (African Americans and whites, Hispanics and whites) and gender (girls and boys). The sampling plan calls for oversampling during the pilot test in order to collect data on each item for 200 African Americans, 168 Hispanics, and 400 girls.

The proposed sample sizes and methods for DIF analyses are acceptable. However, the contractor needs to provide more information about the ways in which DIF data will be analyzed and used. It is important to know what role DIF statistics will play in making judgments about the acceptability of items. Will strict cutoffs based on statistical indices be used to eliminate items? How will human judgment be incorporated into the decision-making process? It is also important to know whether expert reviews for sensitivity are being conducted prior to the pilot testing and whether the experts for such reviews will be consulted in evaluating items flagged for DIF. In addition, what will happen when DIF favors one focal group but disadvantages others (e.g., favors Hispanics but disadvantages African Americans)?

Computation of DIF statistics requires the formulation of a criterion score on which to match individuals. The pilot test form administration plan (which includes the hybrid forms) creates a series of half tests. Half tests are paired to create a full test form, as described above, and items appear in only a single half test. Will the criterion score be formed using the half test in which the item appeared? In this case, the criterion score will be based on smaller numbers of items, which will affect the reliability of the ability estimate. Or will the criterion score be formed using the total score for the two paired half tests so that multiple estimates will exist for each item (e.g., 1a combined with 1b compared with 1a combined with 2b)? Will a two-step process be used that includes refinement of the matching criterion through the elimination of extreme DIF items from the criterion score? What are the plans for dealing with these issues?

RECOMMENDATION 4.3 NAGB and its contractor should continue to detail plans for analyzing the pilot test data. Additional specifications should be provided for assessing the extent to which each item fits the model being used for calibration and the ways in which differential item functioning analyses results will be used in making decisions about the items.

ASSEMBLING FIELD TEST FORMS

The primary purpose of the field test is to try out six operational forms for each of the two subject areas. Field test data will yield information on the psychometric properties of each form, provide the basis for equating the different forms to a common scale, provide normative information on student performance on each test, and also provide data for linking VNT scores to the NAEP achievement levels. The committee considered a number of topics related to plans for the field test, including issues in constructing the operational forms, sampling issues, and plans for analyses.

Targets for item difficulty or, more importantly, test information at different score levels need to be set before operational forms can be assembled. Item difficulty targets speak to the expected difficulty of the test forms. The test information function provides estimates of the expected accuracy of a test form at different score levels. Test difficulty and information functions are developed to be in accordance with the intended purposes and uses of a test. Preliminary information, which lays out the main issues, is available from the contractor (American Institutes for Research, 1999g). Figure 4-1, taken from the document, shows four potential test information functions for the VNT.

As stated above, test information functions indicate the level of precision with which each ability level (or score) is estimated. Tests can be designed to maximize the amount of information provided at specific test scores, usually the test scores of most interest (e.g., where a cutpoint occurs or where most students perform). One of the test information functions (line 3) maximizes test information between the proficient and the advanced score levels. This function, in particular, will not be useful, since the majority of the students are likely to be at the basic score level. Another function (line 4) maximizes information across all score levels. The contractor (American Institutes for Research, 1999l) recently recommended use of a function similar to line 4. However, if the VNT is being constructed from NAEP-like items, there may not be sufficient numbers of easy and hard items in the item pools to provide this level of measurement at the extremes. Use of target test information functions in test assembly will facilitate score comparability across forms and ensure that the measurement properties of the test support the intended score uses.

RECOMMENDATION 4.4 A target test information function should be decided on and set. Although accuracy at all levels is important, accuracy at the lower boundaries of the basic and proficient levels appears most critical. Equivalent accuracy at the lower boundary of the advanced level may not be feasible with the current mix of items, and it may not be desirable because the resulting test would be too difficult for most students.

SPECIAL FORMS FOR BELOW-BASIC AND ADVANCED STUDENTS

The committee is concerned with the difficulty of creating a single test form to provide accurate information about students at both the below-basic and advanced levels. Forms that contain a sufficient number of items to provide accurate information at the advanced level create special problems for students struggling to reach the basic level. The difficult items will be frustrating to these students and may have negative effects on their self-confidence. Furthermore, the difficult items will provide little information on the skill levels for students who score below the basic level. The reverse problem is likely to be true for students at the advanced levels, for whom the easier questions will provide little information.

With NAEP, the focus is on estimating score distributions. Students are not given any feedback on their performance. Questions about which children should be included in the NAEP assessment relate

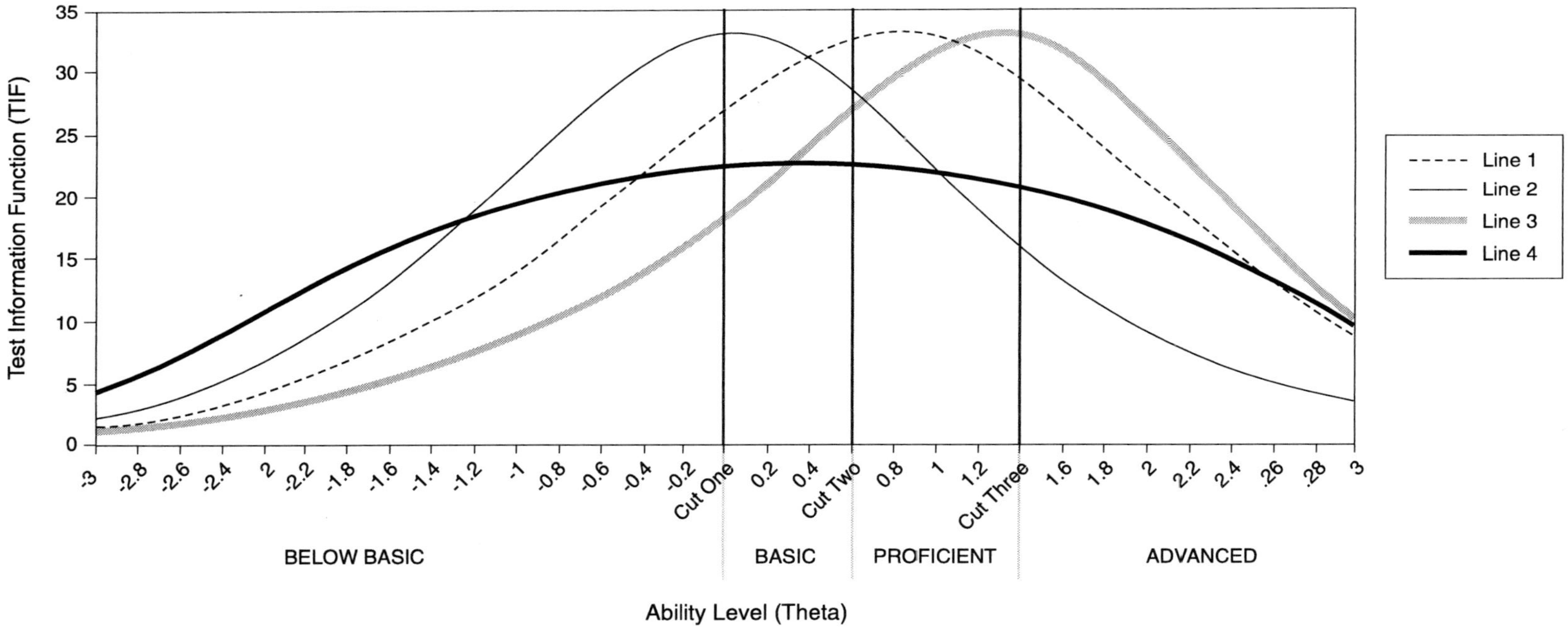

FIGURE 4-1 Hypothetical test information functions. SOURCE: American Institutes for Research (1999g:Fig.1).

NOTES: Line 1 represents a NAEP-like TIF, which has the maximum value around cut two (between basic and proficient).
Line 2 represents a TIF with maximum value at score level lower than NAEP TIF (around cut one, between below basic and basic).
Line 3 represents a TIF with maximum value at score level higher than NAEP TIF (around cut three, between proficient and advanced).
Line 4 represents a TIF with approximately equal precision at all score levels.

to the effects of inclusion on the validity of the score distribution estimates and not to the effects on individual students. The VNT will be quite different in that individual student scores are the primary focus. For this reason, the committee believes it is important to consider the effects of the test scores on the students and parents who receive them. For students classified at the advanced level, there seems little reason for concern. Their interaction with the items should be itself educational, and the feedback they receive will reinforce the value of the skills they have acquired. As indicated above, however, a large proportion of students will not even achieve the basic level. For them, the experience of confronting a large number of questions that they cannot begin to answer will be frustrating at best.

In reviewing reading passages selected for the VNT, the committee noted that all of the passages were written at the 4th-grade level or higher. These items will provide little information on the reading skills of students below the basic achievement level, students who are most in need of feedback about their reading skills. Similarly, students struggling to learn English as a second language may not be able to demonstrate the reading skills they do have unless some easier texts are included. Neither the committee nor NAGB has sufficient information at this time to evaluate how accurately students at different levels will be classified on the basis of their responses to different samples of VNT items. However, it is important that NAGB now consider issues associated with special forms for students judged likely to be below basic, and possibly also for advanced students, so that an informed decision can be made when pilot test data become available. For these reasons, the committee believes it would be desirable to explore the creation of special forms of the VNT for use with students who are likely to find the regular VNT form too difficult. An easier form might also serve as an important accommodation for students with some types of learning disabilities and, especially in the case of the 4th-grade reading test, for students with limited English proficiency. It might also be desirable to consider a special "advanced" form to provide more accurate distinctions at the advanced level.

The need for better information on students below the basic level is a very significant issue for the VNT. As measured by NAEP, in 1998, 38 percent of children in public schools were below the basic level in reading at grade 4; similarly, in 1996, 39 percent were below the basic level in mathematics at grade 8. Among specific populations within the United States, the numbers are much larger: 64 percent of African American students, 60 percent of Hispanic students, and 53 percent of Native American students were below the basic achievement level at grade 4 in reading; also, 45 percent of students in central-city schools and 58 percent of students eligible for free or reduced-price lunches were below the basic level in reading at grade 4 (Donahue et al., 1999). In mathematics, the problem for specific groups was even more dramatic, with 73 percent of African American students, 63 percent of Hispanic students, and 50 percent of Native American students scoring below the basic level in the 1996 grade 8 mathematics assessment; in addition, 53 percent of the students in central-city schools and 72 percent of the Title I participants scored below the basic level (Shaughnessy et al., 1997:93).

In order to promote better performance among the nation's less skilled readers, students, parents, and teachers need high-quality information about their performance. At the moment, many children at grade 4 may not be able to productively participate in the VNT because they will not be able to read the relatively challenging passages on the test. This is especially pertinent for limited-English-proficient students or others with language-based difficulties and can lead to the erroneous conclusion that these children are unable to think about texts or are unable to read anything. In short, there could be no useful information about what they know and can do, a terrible disservice to students who need special attention. Similarly, teachers will not receive information about what these students know and can do, limiting their ability to use information from the VNT to help students and their families.

Unless students and their parents can expect high-quality information from the test, they may question why instructional time should be used for it.

The committee recognizes that there would be added cost and complexity to develop and scale forms of alternate difficulty and to administer such forms appropriately. In addition, explaining the differences between the test forms to test users will be challenging. It may be difficult for users to understand that students taking an easier form are not being held to different standards—a lower standard since they are allowed to take an easier form or a higher standard since they would need a higher percentage of items correct to be classified basic or proficient. Interpretive material would be needed to help students, teachers, and parents understand the information reported and released from the tests (test questions, answers to test questions, percent correct scores, etc.), given that the information would be based on tests of different difficulty levels for different students. These issues commonly arise with all forms of adaptive testing, however, and while complex, have been explained to the public in connection with other testing programs. We recommend, therefore, that exploration of this concept focus initially on the feasibility of creating an easier form to provide more information on students who score below the basic category. Further consideration of the potential benefits and problems associated with multiple-level forms could also be added to ongoing efforts to evaluate alternative reporting options.

Creation of an easy form (in place of one of the current six forms) would require review of item development plans to ensure an adequate supply of easier items in each content and format area. For reading, the issue is not just creating easier items, but also including passages with significantly lower reading levels (i.e., well below the average 4th-grade level). Given current time constraints, we realize that the creation of special easy (or difficult) forms would most likely have to be planned as an addition to the VNT in subsequent years. Yet the potential value of such forms warrants an early consideration of feasibility and potential advantages and disadvantages.

RECOMMENDATION 4.5 NAGB should consider plans for development of an alternate form of the VNT targeted to students at the low end of the achievement scale.

LINKING VNT SCORES TO NAEP ACHIEVEMENT LEVELS

The primary purpose of a link between VNT and NAEP is to enable students to compare their performance with national standards. The content specifications for the VNT are based on the NAEP frameworks, leading to a not unreasonable expectation that scores on the VNT should be able to be compared with the NAEP standards for student achievement on the content covered by these frameworks.

Linking scores from various tests was the subject of an NRC study conducted in 1998. The Committee on Equivalency and Linkage considered the feasibility of developing a scale to compare, or link, scores from existing commercial and state tests to each other and to NAEP. The committee's conclusions were generally negative (National Research Council, 1999e:4-5):

(1) Comparing the full array of currently administered commercial and state achievement tests to one another, through the development of a single equivalency or linking scale, is not feasible.
(2) Reporting individual student scores from the full array of state and commercial achievement tests on the NAEP scale and transforming individual scores on these various tests and assessments into the NAEP achievement levels are not feasible.
(3) Under limited conditions it may be possible to calculate a linkage between two tests, but

multiple factors affect the validity of inferences drawn from the linked scores. These factors include the content, format, and margins of error of the tests; the intended and actual uses of the tests; and the consequences attached to the results of the tests. When tests differ on any of these factors, some limited interpretations of the linked results may be defensible while others would not.

(4) Links between most existing tests and NAEP, for the purposes of reporting individual students' scores on the NAEP scale and in terms of the NAEP achievement levels will be problematic. Unless the test to be linked to NAEP is very similar to NAEP in content, format, and uses, the resulting linkage is likely to be unstable and potentially misleading.

The VNT is designed to be as similar to NAEP as possible and so some of the problems described in the NRC linkage report may be ameliorated. A subsequent study was done by the Linking Feasibility Team (LFT), commissioned by NAGB to investigate the feasibility of linking scores on the proposed VNT and NAEP. The LFT recommended linking through the method of calibration, if the following assumptions are met (Cizek et al., 1999:iii): "1) the VNT and NAEP measure the same constructs as established by content review; 2) the VNT and NAEP measure the NAEP constructs as established empirically; and 3) the content of the VNT can support NAEP achievement levels descriptions." The report states that if these requirements are met, VNT scores could be interpreted directly as estimates of NAEP scores, and NAEP achievement-level descriptions could be used to help interpret VNT achievement-level estimates. The authors conclude that calibration is the only methodology that would lead to such direct interpretation. If these requirements are not met, they recommend the use of social moderation, a judgmental procedure.[1] New labels for achievement levels would be needed and new achievement-level descriptions would be created.

As detailed in the reports from the preceding efforts (National Research Council, 1999e: Cizek et al., 1999), the relationship between VNT and NAEP will be less than perfect for a number of diverse reasons:

- Even though the VNT is based on the NAEP frameworks, the VNT is not being built to exactly mirror the content and statistical specifications of NAEP. For example, the VNT contains a smaller proportion of constructed-response questions in order to make it more efficient to administer and score.
- The VNT is being designed to provide more information than NAEP at the basic and "below basic" achievement levels because VNT scores will be reported for individuals, and a large number of students perform below the basic level.
- VNT scores will be observed scores for individual test takers and, as such, will contain measurement error.
- Different students take different subsets of items during a NAEP assessment so NAEP scores do not exist, and are not reported, for individual test takers. Instead, student data are used to estimate true score distributions for various population subgroups.
- The administration conditions for the two tests will differ. For example, VNT administration may not be centrally monitored in the same way as NAEP. Also, students may be allowed to use

[1]Linkages between the two measures can be obtained by matching distributions of scores and deriving a score-to-score correspondence, using procedures like those used for equating except that no presumption is made that the two tests measure the same variable. Such procedures are called *statistical moderation*. Linkages can also be established when score distribution matching is judgmentally derived; such procedures are referred to as *social moderation* (Cizek et al., 1999).

their own calculators on the mathematics test instead of using standard calculators, as is done with NAEP.

- Because the VNT will report scores to students, parents, and teachers, there is the potential for much higher motivation for students taking the VNT.

The committee agrees with the findings of the LFT (Cizek et al., 1999) and others that, because of the differences in test specifications, scoring, and administration conditions, it will not be possible to create a strong link between NAEP and VNT scores, regardless of the linking procedure used. As a result, it will be inappropriate for educators and policy makers to make inferences about a group's performance on NAEP based on data from the VNT.

The NAEP-VNT linkage is different from many of the examples of failed linkages (National Research Council, 1999e) in the high degree of content similarity between the two exams. There is no precise measure of content similarity or its effect on linkages, and the LFT identified a number of ways in which the format and content of the VNT will differ from NAEP. So far, we have only expert judgment on the extent to which prior research will generalize to VNT-NAEP linkage efforts and on the amount of error, initially and over time, that will be introduced by the differences in administration and use between the tests.

Although it may not be possible to create a linkage between NAEP and VNT that permits direct inferences from one to the other, it may be possible to establish a link that supports other types of inferences. For example, suppose a linkage could be created that permits inferences from student performance on the VNT to student performance on NAEP when NAEP is given under VNT conditions.

To create this type of linkage, a short form of NAEP (representative of the content and statistical specifications for NAEP) would be constructed and spiraled along with the VNT under VNT conditions (e.g., two 45-minute sections, students provide their own calculators). The linkage between the VNT and the NAEP-like VNT (or short-form of NAEP) could then be accomplished through a simultaneous IRT scaling of the VNT and short-form NAEP items, through separate IRT scalings followed by a Stocking-Lord test characteristic function transformation, or through an equipercentile matching of the raw score distributions.

Achievement-level cutpoints on the short-form NAEP scale could be obtained by using the judgmental proportions correct from the achievement-level setting used for the main NAEP. These Angoff proportions would be projected onto the short-form NAEP scale using the same procedure as used by ACT to project the Angoff proportions on the main NAEP scale. These cutpoints would then be projected to the VNT scale by using the linkage established between VNT and NAEP given under VNT administration conditions.

Item parameters computed for the short-form NAEP items administered under main NAEP conditions could be compared with those computed from data collected under VNT conditions in order to check the comparability of the achievement-level cutpoints between the main NAEP and the short-form NAEP. Other checks on the quality of the linkage between VNT and short-form NAEP would include the stability of the linkage function for a variety of subgroups (e.g., males and females) and similarity in the shapes of the proficiency distributions for the two types of tests.

The above plan is very similar to one proposed by Johnson (Johnson,1999a). It differs only in that the test to which the VNT is being linked is built from NAEP items to be as representative of NAEP as possible, rather than to be as representative of the VNT as possible. This approach may make the observed linkage slightly less stable because of potentially greater content differences between the tests being linked. But it does strengthen the extent to which the resulting linkage can be interpreted in terms of student performance on NAEP when NAEP is given under VNT conditions. Furthermore, it increases the perceived face validity of the linkage.

We believe that the above linkage plan should be tried out during the pilot test. At a minimum, it would require the development of a two-section short-form NAEP. The short-form NAEP would be spiraled with the VNT forms in the pilot test in at least one school cluster. In addition, hybrid forms composed of one VNT section and one NAEP section could be included to reduce the need to assume equivalence of samples within a school cluster. Ideally, more than one short-form NAEP would be developed and included in the pilot test, which would enable the resulting linkage to take account of form-to-form differences.

Because the main NAEP, the short-form NAEP, and the VNT would each be based on the same content frameworks, use similar item types, and be administered to the same population, we recommend that the linkage be based on an empirical procedure rather than on social moderation. We would only recommend the use of social moderation if sample sizes were too small to support an empirical linkage.

Investigation of the linkage during the pilot test would provide valuable information that could be used to refine the actual linkage plan to be carried out during the field test. It would also provide useful information about the anticipated quality of the linkage. Such information would be helpful in preparing guidelines for score use, score reports, and score interpretative materials, including those related to score aggregation. It would also expedite the actual score reporting process following the field test.

RECOMMENDATION 4.6 Plans for the VNT pilot test should include efforts to gather empirical data on the effects of content, administration, and use differences between the VNT and NAEP on the feasibility of linking VNT scores to the NAEP score scale. Specifically, a NAEP-like form (e.g., two non-overlapping booklets from recent 4th-grade reading and 8th-grade mathematics assessments) should be included to allow for an assessment of the effect of content differences and administration differences on the linkage of VNT scores to the NAEP scale.

5

Inclusion and Accommodation

In the November 1997 legislation that established the National Assessment Governing Board's responsibility for the development of the Voluntary National Tests, Congress required NAGB to make four determinations. The third of these is "whether the test development process and test items take into account the needs of disadvantaged, limited English proficient and disabled students" (P.L. 105-78:Sec. 307 (b)(3)). The same legislation called on the National Research Council "to evaluate whether the test items address the needs of disadvantaged, limited English proficient, and disabled students."

There are two key challenges to testing students with disabilities or limited English proficiency. The first is to establish effective procedures for identifying and screening such students, so they can appropriately be included in assessment programs. Federal law and state and local policy increasingly demand participation of these special populations in all education activities, both as a means of establishing the needs and progress of individual students and for purposes of system accountability. The second challenge is to identify and provide necessary accommodations (e.g., large-print type, extended time) to students with special needs while maintaining comparable test validity with that for the general population (see National Research Council, 1997, 1999e). That is, any accommodation should alter only the conditions of assessment without otherwise affecting the measurement of performance.

This issue is growing in importance, as is the number of students with disabilities or with limited English proficiency. Students with disabilities are now 12.3 percent of all students in elementary and secondary schools, and students with limited English proficiency are 5.5 percent of all students. For the latter category, however, it is important to remember that in some local jurisdictions the percentage may be as high as 60 percent. There is a need to be responsive and inclusive so that these increasing numbers of students can be optimally served by the nation's educational system. We concur with the observation in the earlier NRC report (National Research Council, 1999b:46): "The federal government has an important leadership role to play in subsidizing and demonstrating valid efforts to include these populations."

In this chapter we first review the issue of inclusion and accommodation in the VNT, along with the findings and recommendations of the NRC on the year 1 evaluation of VNT development. We then review and assess the treatment of inclusion and accommodation in VNT development activities from fall 1998 through the August 1999 NAGB meeting.

To accomplish these objectives, the committee reviewed the following documents:

- Voluntary National Test: Inclusions and Accommodations for Test Development. Policy Statement (National Assessment Governing Board, 1998a)
- Increasing Participation of Special Needs Students in NAEP: Results of the 1996 Research Study (Mazzeo, 1999)
- Cognitive Lab Report (American Institutes for Research, 1998d)
- Background Paper Reviewing Laws and Regulations, Current Practice, and Research Relevant to Inclusion and Accommodations for Students with Disabilities (American Institutes for Research, 1998b)
- Revised Inclusion and Accommodations Workplan (American Institutes for Research, 1998k)
- Summary of VNT Activities and Plans Related to the Assessment of Students with Disabilities and Students with Limited English Proficiency (American Institutes for Research, 1998m)
- List of Groups and Agencies with Interests in Issues Related to Inclusion and Accommodations for Students with Disabilities (American Institutes for Research, 1998h)
- List of Groups and Agencies with Interests in Issues Related to Inclusion and Accommodations for Students with Limited English Proficiency (American Institutes for Research, 1998i)
- *Evaluation of the Voluntary National Tests, Phase I Report* (National Research Council, 1999b)
- *High Stakes: Testing for Tracking, Promotion, and Graduation* (National Research Council, 1999d)
- Cognitive Lab Report: Lessons Learned (American Institutes for Research, 1999a)
- VNT: Issues Concerning Score Reporting for the Voluntary National Tests: Results of Parent and Teacher Focus Groups (American Institutes for Research, 1999p)
- Background Paper Reviewing Laws and Regulations, Current Practice, and Research Relevant to Inclusion and Accommodations for Students with Disabilities (American Institutes for Research, 1998a)
- Background Paper Reviewing Laws and Regulations, Current Practice, and Research Relevant to Inclusion and Accommodations for Students with Limited English Proficiency (American Institutes for Research, 1998c)
- Public Hearings and Written Testimony on Students with Disabilities and the Proposed Voluntary National Test: October-November 1998. Synthesis Report (National Assessment Governing Board, 1999b)
- Public Hearings and Written Testimony on Students with Limited English Proficiency and the Proposed Voluntary National Test: October-November 1998. Synthesis Report (National Assessment Governing Board, 1999c)
- Effects of Extended Time and Small-Group-Administration Accommodations Study 2 in Proposed Year 3-5 Research Plan (American Institutes for Research, 1999e)
- *Improving Schooling for Language-Minority Children: A Research Agenda* (National Research Council and Institute of Medicine, 1997)
- *Preventing Reading Difficulties in Young Children* (National Research Council, 1998b).

NAGB AND AIR ACTIVITIES

Initial Plans

In summer 1998 NAGB approved a policy for the inclusion and accommodation of students with disabilities and limited English proficiency in the pilot test of the VNT (National Assessment Governing Board, 1998a). The policy was based on four general principles: (1) that the current NAEP inclusion criteria would be used with the VNT pilot test, (2) that so far as feasible the testing accommodations used with NAEP would be provided to students who needed them in the VNT pilot, (3) that decisions to accommodate individual students would be made by knowledgeable school staff in consultation with the pilot test examiner, and (4) that records would be kept of any accommodations used by each student. The policy stated the inclusion criteria and listed the allowable accommodations for students with disabilities and for students with limited English proficiency at each grade level. NAGB revised this policy at its August 1999 meeting to include dual-language booklets (with questions in both English and Spanish) for the 8th-grade mathematics test.

The policy adopted for the pilot test is based on research conducted in conjunction with the 1996 NAEP in mathematics and science. From its beginning around 1970 through the middle 1990s, NAEP assessments had been carried out without any kind of accommodation. Procedures for "inclusion" usually focused on the exclusion of some students from the assessments, rather than on universal participation. Since the mid-1990s, as the growth of special student populations and the importance of their participation in large-scale assessments have increasingly been recognized, NAEP has experimented with new, more inclusive participation and accommodation policies which are reflected in the NAGB principles for the VNT and the AIR planning documents.

In 1996, there were essentially two "experimental" conditions in NAEP in addition to the then operational sample (S1): a sample of schools where the inclusion criteria were changed but no new accommodations were offered (S2) and a sample of schools where the new criteria were used and new accommodations (extra time, small-group administration, bilingual booklets, and large-print or Braille booklets) were offered (S3). The change in the inclusion criterion for students with disabilities had little effect on participation (slightly fewer grade 4 students participated). The change in inclusion criteria for limited-English-proficient students resulted in a significant (20%) drop in the participation of grade 4 students, but did not affect participation at grades 8 or 12. Comparison of the S2 and S3 samples showed that the additional accommodations did significantly increase participation, particularly for Spanish-language students (Mazzeo, presentation to NAGB, August 7, 1999).[1]

NAGB commissioned background reports from AIR on the inclusion and accommodation of students with disabilities and with limited English proficiency, but as of summer 1998 there had been little developmental work to address the special needs of these populations. The 584 participants in cognitive laboratory sessions included students with disabilities (32 in reading and 19 in math) and limited English proficiency (11 in reading and 12 in math), but their numbers were too small to provide substantial or reliable information about the potential participation of such students in the VNT (American Institutes for Research, 1998d). In a later document, American Institutes for Research (1998m:2) reports that "29 of the reading participants and 18 of the mathematics participants had some sort of special educational need other than gifted and talented. No systematic data were gathered, however, on the numbers of these students who were receiving special educational services guided by an individualized educational plan."

[1]A more complete report of the results of this research is scheduled for release by the National Center for Education Statistics in fall 1999.

All of the other contractor documents (American Institutes for Research, 1998b, 1998h, 1998i, 1998k) involve proposals, but no specific development activities or plans aimed at the needs of disadvantaged students or those with limited English proficiency had been undertaken.

NAGB's draft principles (labeled as a policy statement) reflect the NAEP experiments: depending on the test and population in question, accommodations for a pilot test may include large-print booklets, extended time, small-group administration, one-on-one administration, a scribe or computer to record answers, reading a test aloud by an examiner, other format or equipment modifications, or a bilingual dictionary if it is normally allowed by the school. However, the success of these policies in increasing participation has not yet been established (see above), nor have their effects on test performance and score comparability been validated.

Year 2 Plans

Since fall 1998, NAGB and AIR have continued information-gathering and planning activities related to inclusion and accommodation in the VNT. Although progress has been made, the committee finds that these activities have not fully implemented the earlier NRC recommendation for accelerated work on inclusion and accommodation.

In its November 1998 report, American Institutes for Research (1998m) reviewed steps taken during item development to consider the needs of special students. They included passage review, plain language review, item editing, bias and sensitivity review, cognitive lab participation, and item analysis. Although these steps are appropriate and commendable, they do not focus specifically on the needs of students with disabilities or with limited English proficiency. We have already noted the small number of disabled students and English-language learners in the first year of cognitive laboratories. In fact, AIR's March 1999 report (American Institutes for Research, 1999a) on lessons learned in the first-year cognitive laboratories makes no reference to inclusion or accommodation issues or effects. Earlier, in commenting on plans for item analysis, American Institutes for Research (1998m:2) reported that, "even with the planned oversampling of Hispanic students [in the pilot test], it was not anticipated that sufficient numbers of either students with disabilities or students with limited English proficiency would have been included to permit DIF (differential item functioning) analysis based on these characteristics."

AIR also suggested two enhancements of item development activities during year 2 (American Institutes for Research, 1998m:3): (1) "to review items against plain language guidelines that have been developed under the sponsorship of the Goals 2000 program and elsewhere" and (2) to include larger numbers of students with disabilities and limited English proficiency in the second year of cognitive laboratories. The committee has been informed that both of these proposals have been adopted. In particular, an effort is being made to include students with disabilities and with limited English proficiency among the nine students in each item trial in this year's cognitive laboratories.

The AIR planning document also reports that, in connection with its exploration of reporting and test use, focus groups with parents and teachers included (American Institutes for Research, 1998m:5): "one composed of special education teachers, one comprising parents of special education students, and one involving teachers of students with limited English proficiency." A report of the findings of these focus groups was completed in April of this year (American Institutes for Research, 1999p). It consisted of a general description of parents' and teachers' comments regarding draft score reporting materials, additional desired score reporting materials, information desired by teachers to help them increase the academic achievement levels of their students, and ways teachers could communicate with parents about increasing student achievement. However, no mention was made of the special needs of students with disabilities and limited English proficient students in reporting proposed VNT score information.

This lack of attention to the problems of inclusion and accommodation is somewhat puzzling in the light of AIR's background documents (American Institutes for Research, 1998a, 1998b, 1998c). In the committee's judgment, those three reports are thorough, competent, and cogent, and they would have provided an ample basis for additional development activities to address the needs of students with disabilities and with limited English proficiency. An AIR planning document (American Institutes for Research, 1998m:5) also acknowledged the issue:

> Following the release of the NRC report on appropriate test use in September 1998, the board undertook a series of public hearings in which it invited citizens concerned with the education and assessment of students with disabilities or limited English proficiency to comment on the findings of the NRC report and the evolving plans for addressing the needs of these students in the VNT.

Invitations to the hearings were developed on the basis of lists of relevant groups and agencies that were assembled by AIR (American Institutes for Research, 1998h, 1998i). The hearings were held during October and November 1998 in several large cities: Washington, DC; Atlanta, GA; New York, NY; Chicago, IL; Austin, TX; and Los Angeles, CA. Witnesses were asked to respond to a series of questions about inclusion and accommodation in large-scale assessment, including the proposed VNT. For example, in the case of students with disabilities (National Assessment Governing Board 1999b:1-3), there were eight queries:

1. What should be the criteria for including a student with disabilities in standard administrations of large-scale tests?
2. What, if any, should be the criteria for exempting a student with disabilities from standard administrations of large-scale tests? What examples exist of such criteria that have been empirically determined and/or validated?
3. What accommodations in the administration of the test should be considered for students with disabilities and according to what criteria should those accommodations be provided? What adaptations (i.e., changes or special versions) to the test should be considered and under what circumstances should they be provided?
4. What validation evidence is sufficient to conclude that results from testing with accommodations/adaptations are comparable to results from testing without accommodations/adaptations? Until such validity is established, what cautions about interpretation, if any, should accompany individual test results?
5. The Voluntary National Test is intended to provide individual student results. What are the pros and cons of permitting aggregate results to be reported for students with disabilities within a school, district, or state?
6. In setting performance standards on a test for interpreting test results, what, if any, considerations should be given with regard to implications for students with disabilities?
7. What specific criteria should be considered in reviewing test items as they are being developed to optimize the inclusion of students with disabilities? What other specific considerations should be taken into account in test development?
8. What other issues related to inclusion and accommodations in testing for students with disabilities should be considered by the Governing Board for the Voluntary National Tests in order to ensure fair and equitable test administration for all students?

Parallel questions were also posed with reference to students with limited English proficiency (see National Assessment Governing Board, 1999e). In addition, witnesses who addressed the testing of students with limited English proficiency were asked: "What are the technical and policy pros and cons of testing students with limited English proficiency in the student's native language and under what circumstance, if any, might this be considered as an appropriate adaptation in large-scale testing?" (National Assessment Governing Board, 1999c:3-5).

The testimony at these two series of hearings was recorded and summarized in two reports (National Assessment Governing Board, 1999b, 1999c). While the hearings may have been useful in providing a platform for a number of interested groups and agencies—and signaling the interest of NAGB in the views of those groups and agencies—the committee does not find that the hearing summaries add substantial new information or ideas that could affect the design of the VNT.

AIR has proposed a study of the effect of two particular accommodations—extended time and small-group administration—on VNT performance, which would be carried out in connection with the pilot test (American Institutes for Research, 1999e). These two accommodations are common, and, thus, the committee considers it important to understand their effects on test performance and validity. This proposal appeared in a proposed year 2 research plan, but that plan was abandoned because of the congressional ban on pilot testing during 1999. The proposal was modified for year 3 of development, in connection with the year 2000 pilot test (American Institutes for Research, 1999e:8-9), which states:

> It is important to determine the impact that extra time has on student performance and whether only examinees with disabilities or limited English proficiency benefit from having extra time, or the degree to which the scores of other examinees also would improve with such accommodations.

In the case of small-group administration, the research proposal focuses on students with individual education plans (IEPs) and is designed to estimate the effects of extended time and small-group administration, but is not designed to compare the performance of students with IEPs to that of other students under the experimental conditions.

Each study will be based on a single pilot test form. The first, extended-time study is proposed for a supplementary sample of large schools in order to minimize the cost of obtaining 150 observations in each cell of the design. The experimental design crosses the three student groups by two times of administration: standard time (45 minutes) and twice the standard time (90 minutes). The second study will use a combination of schools in the main pilot sample and a supplementary sample of schools. There are five cells, crossing standard administration, embedded small groups, and pull-out small groups with standard time limits and double time limits (excluding the double-time condition with standard administration). In both studies, it should be possible to assess the speededness of the tests, as well as differences in performance between groups of students and experimental conditions. The analysis of the first study is also proposed to include analyses of differential item functioning among students with disabilities, students with limited English proficiency, and students who are not in these special groups. Since there are only 300 students in each of the three groups, the committee is concerned that these analyses may not have sufficient statistical power, relative to the design of the main pilot sample. However, the committee does agree that it will be valuable, first, to assess the influence of these two accommodations on VNT performance.

CONCLUSIONS AND RECOMMENDATIONS

Year 1 Report

In its year 1 VNT evaluation report, the National Research Council (1999b:46) noted:

> Because of the compressed schedule in the early phases of VNT development, along with the desire to achieve close correspondence between the VNT and NAEP, the NAGB plans and the AIR background paper on students with disabilities both focus on recent NAEP practices for inclusion and accommodation, rather than taking a broader, more proactive stance. We believe the federal government has an important leadership role to play in subsidizing and demonstrating valid efforts to include these populations . . . . The procedures discussed in the draft documents are intended to increase participation and provide valid assessments for all students, but they essentially involve retrofitting established assessment instruments and procedures to special populations of students; another approach would be to design and develop assessments from the beginning that are accessible to and valid for all students . . . .
>
> Unless extensive development work is done with students with disabilities and with limited English proficiency, it would be unreasonable to expect that the VNT will be valid for use with these student populations. Both of these populations are heterogeneous, e.g., in primary language, level of proficiency in English, and specific type of disability. Moreover, they differ from the majority of students, not only in ways that affect test-taking directly—e.g., those that can be accommodated through additional time or assistive devices—but also in styles of learning and in levels of motivation or anxiety. Such differences are very likely to reduce the validity and comparability of test performance.

The Committee on Appropriate Test Use identified two important ways in which inclusion and accommodation can be improved (National Research Council, 1999d). First, the focus should be on inclusion and accommodation issues throughout item and test development, so a test is designed from the ground up to be accessible and comparable among special populations. For example, oversampling of students with disabilities and with limited English proficiency in the course of pilot testing will provide sufficient numbers of cases in major subgroups of these students to permit key statistical analyses. Second, test developers should explore the use of new technologies—such as computer-based adaptive testing for students who need extra time—that show promise for substantially reducing or eliminating irrelevant performance differentials between students who require accommodation and other students. The report recognized, however, that development work of this kind is just beginning, and there are presently few exemplars of it.

The NRC year 1 evaluation report concluded (National Research Council, 1999b:47):

> The statement of principles and the AIR planning documents provide a limited basis for evaluation of provisions for inclusion and accommodation in the VNT—and no specific basis to address the quality of item development relative to the needs of those students . . . . A major opportunity for improved large-scale assessment is being lost in NAGB's conservative approach to inclusion and accommodation in the VNT.

It thus recommended and explained (National Research Council, 1999b:47):

> **NAGB should accelerate its plans and schedule for inclusion and accommodation of students with disabilities and limited English proficiency in order to increase the participation of both those student populations and to increase the comparability of VNT performance among student populations . . . .**
>
> This recommendation requires prompt action because so much of the development work in the first round of the VNT has already been completed. We have already noted the modest attention to students with special needs in the cognitive laboratory sessions. In the pilot test, NAGB plans to identify students with disabilities and with limited English proficiency and with the types of accommodations that have been provided. However, there are no provisions in the design to ensure that there will be sufficient numbers of these students—such as students requiring specific types of accommodation—to support reliable DIF analyses. We think that it would be feasible to include

larger numbers of such students in the pilot and field tests, for example, by increasing sampling fractions of such students within schools. Moreover, there appears to be no plan to translate the 8th-grade mathematics test into Spanish (or any other language), a decision that is likely to affect participation in the VNT by major school districts. There has been some discussion of a Spanish translation after the field test, but this would be too late for the item analyses needed to construct comparable English and Spanish forms.

Next Steps

After release of the NRC report on year 1 of VNT development, NAGB completed several activities related to inclusion and accommodation of students with disabilities and with limited English proficiency. They included preparation of two background papers, public hearings, and the summary of testimony from those hearings. However, these activities have not yet carried the development process much beyond its state in the summer of 1998, when the report found (National Research Council, 1999b:47) "a limited basis for evaluation of provisions for inclusion and accommodation in the VNT—and no specific basis to address the quality of item development relative to the needs of those students." We applaud AIR's proposal to evaluate the effects of two common accommodations on VNT performance among students with disabilities and with limited English proficiency in the pilot test.

Consistent with recommendations of both the NRC year 1 evaluation report and the committee's interim report, plans for the pilot test were changed to include provision of a dual-language translation of the 8th-grade mathematics test. In addition, proposed research on extra time and small-group or one-on-one administration in conjunction with the pilot test were approved at the August 1999 NAGB meeting. With respect to the provision of dual-language booklets, while language simplification methodology has been used in the test development process, little attention has been paid to other language issues regarding the VNT mathematics test (e.g., whether translation of existing questions into Spanish and other languages will produce comparable items or whether there are methods to reduce the reading level of mathematics items). Participation in the cognitive laboratories by students with disabilities and with limited English proficiency has been expanded, and the committee will be interested to learn how this information will be used in item and test development.

The committee is concerned about implementation of the principles of inclusion and accommodation that were adopted by the National Assessment Governing Board (1998a) for the pilot test. Given the modest subsequent development activity, the committee can only assume that these principles are likely to be applied in the field test and under operational conditions. However, the committee is concerned that it may be more difficult to provide these accommodations on a large scale than under the operational conditions of NAEP, where they are now standard procedure.

RECOMMENDATION 5.1 NAGB should accelerate its plans, research, and schedule for inclusion and accommodation of students with disabilities and limited English proficiency in order to increase the participation of both those student populations in numbers representative of their numbers in the student population.

Although the condensed schedule of test development during year 1 provided a substantial rationale for limited attention to issues of inclusion and accommodation, the extension of the development schedule for another year prior to pilot testing provided extra time for more intensive consideration of these issues. That window is more than half closed, and there are modest signs of progress: the effort to include more students in cognitive labs, the intent to apply guidelines for simplified language, and the AIR proposal to study effects of accommodation in the pilot test.

We offer additional recommendations to underscore our concern for the urgency and NAGB's need to expand inclusion and accommodation efforts for the VNT.

RECOMMENDATION 5.2 NAGB should consider expanding the accommodation research planned in conjunction with the pilot test to include a systematic analysis of the use and effect of dual-language booklets. Additional accommodations for English-language learners, in the forms of both a Spanish-only translation of the mathematics test and the use of English-Spanish and English-other languages dictionaries for the mathematics test, should also be considered for the pilot test.

Because of the large number of limited-English-proficient students in some areas and the fact that a large percentage of those students speak Spanish, the committee applauds NAGB's decision to provide a side-by-side Spanish translation of the 8th-grade mathematics test. Providing this dual-language testing accommodation and the related accommodations we propose in Recommendation 5.2 will allow more limited-English-proficient students to demonstrate what they know and can do in mathematics.

In addition it is important not to lose sight of related VNT longer term issues and needs. For example, while perhaps not feasible in the pilot test, the concern to build tests "from the ground up" rather than translate them after they have been developed in English is important to consider for the field test. Certainly more direct approaches to second-language item development should not be ruled out for test development purposes. There also needs to be an ongoing research plan beyond the pilot test that continually promises to advance the inclusion objective for students with disabilities and who have limited English proficiency.

The issue of the VNT 4th-grade reading test is clearly a sensitive one. On one hand are the existing NAEP guidelines and tradition regarding the English-only nature of this measure. On the other hand is the desire to include as many students as possible and to provide necessary and reasonable accommodations to accomplish inclusion. Thus, it is important as a basic premise that NAGB clearly articulate the constructs to be measured by the VNT 4th-grade reading test so that reasonable accommodations can be considered to fulfill the goal of maximum inclusion.

RECOMMENDATION 5.3 NAGB should clarify the reading constructs (e.g., reading proficiency, reading proficiency in English, etc.) being measured by the 4th-grade reading test prior to the field test and then address what accommodations would not invalidate assessment of these constructs. In particular, NAGB should clarify when reading competency could be assessed in a student's primary or native language if it is not English.

This issue needs to be addressed in the very near future. It is especially important in light of the grade level of the reading test (4th), the current NAEP rules for inclusion of English-language learners on the basis of years of English-reading school experience (3 years), and the weight of the evidence as shown in two recent reports (National Research Council and Institute of Medicine, 1997; National Research Council, 1998b) that document the greater likelihood of students developing strong English literacy skills if they first learn to read and write in their primary (native) language.

There remains a critical need to obtain statistical information on the efficacy of different accommodation for students with disabilities and those who have limited English proficiency. Some studies and analyses will likely require oversampling of students within specific special populations. This should begin with the pilot test version as planned.

RECOMMENDATION 5.4 The National Assessment Governing Board should assess the effects of various accommodations for limited-English-proficient students and students with disabilities at both the item and total test score levels. To do so will require oversampling in the pilot and field tests.

Finally, it is important for NAGB to provide a clear and concise list of accommodations for administrative use with the VNT, and they should be conveyed in user-friendly language. Given that VNT and NAEP scores may not be comparable, it might not be necessary to strictly emulate the standard NAEP accommodations. Additionally, the role of individualized educational plans should be defined within the context of these suggested VNT accommodations. It is particularly important that the procedures intended for operational administration of the VNT be in place for the field test, as data from the field test will provide both normative information for use in reporting operational results and the basis for linking performance on the VNT test forms to the NAEP achievement levels.

RECOMMENDATION 5.5 The National Assessment Governing Board should provide a clear, concise, and detailed list of accommodations for the VNT for students with disabilities or limited English proficiency for use on the VNT field test.

We think these issues should be addressed in a systematic manner—one that will gather valid, scientific evidence that can be used to improve the inclusiveness and validity of testing students with learning disabilities and students with limited English proficiency. Our specific recommendations suggest a step-by-step approach to learning more about testing these groups. They should provide new evidence that will improve testing guidelines or suggest additional lines of research to improve practice.

6

Reporting

In its statement on the purposes and uses of the VNT, NAGB responded to the congressional request that it include "a description of the achievement levels and reporting methods to be used in grading any national test." Given that the stated purpose of the VNT is to measure individual student achievement and the stated use is to provide information describing the achievement of individual students, NAGB offered the following statements on reporting VNT results (National Assessment Governing Board, 1999e:11):

> . . . results of the voluntary national tests [should] be provided separately for each student. Parents, students, and authorized educators . . . should receive the test results report for the student. Test results should be reported according to performance standards for [NAEP]. These are the NAEP achievement levels: Basic, Proficient, and Advanced. All test questions, student answers, and an answer key should be returned with the test results; it will be clear which questions were answered correctly and which were not. The achievement levels should be explained and illustrated in light of the questions on the test. Also, based on the nationally representative sample of students who participated in the national tryout of the test the year before, the percent of students nationally at each achievement level should be provided with the report.

This chapter considers the *process* leading to NAGB's statements to Congress regarding the reporting of VNT results, and it evaluates the stated *plans* for reporting results, as outlined in the document submitted to Congress. The committee reviewed reporting procedures with two criteria in mind: (1) Would the current plans result in formats accessible to parents, teachers, and students? (2) Would the current plans report results using meaningful metrics?

To accomplish this, the committee reviewed the following documents:

- The Voluntary National Test: Purpose, Intended Use, Definition of Voluntary and Reporting (National Assessment Governing Board, 1999e)
- Overview: Determining the Purpose, Intended Use, Definition of the Term Voluntary and Reporting for the Proposed Voluntary National Test (National Assessment Governing Board, 1999a)

- VNT: Issues Concerning Score Reporting for the Voluntary National Tests: Results of Parent and Teacher Focus Groups, (American Institutes for Research, 1999p)
- Selected Item Response Theory Scoring Options for Estimating Trait Values (from Wendy Yen to American Institutes for Research, 1999h)
- Score Reporting, Scoring Examinees, and Technical Specifications: How Should These be Influenced by the Purposes and Intended Uses of the VNT (American Institutes for Research, 1999g)
- VNT: Plans for Continuing Work in Score Reporting (American Institutes for Research, 1999q)
- VNT: Proposed Score Reporting Metrics and Examinee Scoring Algorithms for the Voluntary National Tests (American Institutes for Research, 1999r)

SCORE COMPUTATION

One of the primary recommendations of the NRC's year 1 report was that decisions about how scores will be computed and reported should be made before the design of the VNT test forms can be fully evaluated (National Research Council, 1999b:51). NAGB and AIR are developing and evaluating options for test use and have conducted focus groups that include consideration of reporting options, but no further decisions about score computation and reporting have been made. We believe that decisions are needed soon.

Three options for computing student scores were identified in a paper prepared for the VNT developer's technical advisory committee (American Institutes for Research, 1999h). One option is pattern scoring, which assigns an overall score to each possible pattern of item scores, based on a model that relates examinee ability to the probability of achieving each observed item score. With pattern scoring, a student's score depends on the particular pattern of right and wrong answers. As a result, individuals with the same number of correct responses may get different scores, depending on *which* items were answered correctly. Conversion of response strings to reported scores is complicated, and it is not likely to be well understood nor accepted by parents, teachers, and others who will have access to both item and total scores.

Another scoring approach is to use nonoptimal weights, with the weights determined according to either judgmental or statistical criteria. For example, easy items can be given less weight than more difficult items, multiple-choice items can be given less weight than open-ended items, or all items can be weighted in proportion to the item's correlation with a total score. Use of such a scoring procedure would make conversion from the number correct to the reported score less complex than with pattern scoring, but it is not as straightforward as a raw score approach.

The raw score approach is the most straightforward method: a total score is determined directly by summing the scores for all the individual items. Multiple-choice items are scored 1 for a correct answer and 0 for an incorrect answer or for no answer. Open-ended response items are scored on a 2-, 3-, 4-, or 5-point scale according to a defined scoring rubric. The particular subset of items that are responded to correctly—whether easier or harder, multiple-choice, or open-ended—is not relevant. We agree with Yen's conclusion (American Institutes for Research, 1999h) that for tests with adequate numbers of items, different item weighting schemes have little effect on the reliability or validity of the total score. Given that test items and answer sheets will be available to students, parents, and teachers, we believe that this straightforward approach makes the most sense.

RECOMMENDATION 6.1 Given that test items and answer sheets will be provided to students, parents, and teachers, as well as made available to the general public, test forms

should be designed to support scoring using a straightforward, total correct, raw score approach.

The committee also suggests that careful thought be given to the manner of awarding points for constructed-response (open-ended) items. The final scaled score for a student will be based on the student's raw score, which is simply the number of points awarded. Since each correct response to a multiple-choice item is worth 1 point, care must be taken to ensure that each partial-credit point awarded for a response to a constructed-response item represents a correct portion of a response. The idea is that a response may need to include several correct assertions in order to be fully correct, so each of those assertions can receive a point. For example, the scoring rubric proposed for one mathematics item that asked students to draw a particular type of geometric figure called for 1 point for an incorrect drawing (with no additional constraints) and 2 points for a correct drawing. This might more appropriately be a 1-point item. Or, since the drawn figure had to satisfy two conditions, it might be appropriate to award 1 point for a figure satisfying one of the conditions and 2 points for a response satisfying both.

In VNT: Proposed Score Reporting Metrics and Examinee Scoring Algorithms for the Voluntary National Tests (American Institutes for Research, 1999r), AIR notes a NAGB decision that "all VNT scoring rubrics for constructed-response items award a score of '1' only to those responses that are at least partially correct"; in contrast, NAEP awards a score of 1 to attempts at providing relevant responses. The committee agrees with this decision. Failure to consider the manner for awarding points could have a serious negative effect on the perceived validity of the scoring process among students, parents, and the public. Awarding points for erroneous work or having too large a disparity in the value of a point given for different responses will be noticed and will be hard to defend.

RECOMMENDATION 6.2 Special attention should be given to the work required for receiving partial credit for constructed-response items that have full scores of more than 1 point.

REPORTING SCALE

For NAEP, the current practice is to summarize performance in terms of achievement levels (basic, proficient, and advanced) by reporting the percentages of students scoring at each achievement level. To remain consistent with NAEP, the VNT will also report scores using these achievement-level categories. One shortcoming of this type of reporting is that it does not show where a student placed within the category. For instance, was the student close to the upper boundary of basic and nearly in the proficient category? Or, did he or she only just make it over the lower boundary into the basic category? This shortcoming can be overcome if achievement-level reporting is supplemented with reporting using a standardized numeric scale, such as a scale score. This manner of reporting will show how close a student is to the next higher or next lower achievement-level boundary. This additional information will be particularly important in settings where a significant portion of the students score in the "below basic" category.

To ensure clear understanding by students, parents, and teachers, the released materials could include a table that allows mapping of the possible scores to specific points on the standardized scale. A separate table would be needed for each VNT form to account for minor differences in the difficulty of each form. In addition, the cutpoints for each achievement level could be fixed and prominently displayed on the standardized numeric scale.

Tests that report performance according to achievement-level categories might include a series of probabilistic statements regarding achievement-level classification errors. For example, a score report might say that of the students who performed at the basic level, 68 percent would actually be at the basic level of achievement, 23 percent at the proficient level, and 5 percent at the advanced level. Although provision of information regarding the precision with which performance is measured is essential, we do not believe these sorts of probabilistic statements are particularly helpful for parents and do not recommend their use for the VNT.

In conducting focus groups on reporting issues with parents and teachers, NAGB and AIR have developed a potential reporting format that combines NAEP achievement levels with a continuous score scale. In this format, error is communicated by a band around a point estimate on the underlying continuous scale. Reporting of confidence band information will allow parents to see the precision with which their children are placed into the various achievement levels without having to grapple with classification error probabilities. We believe that parents will find information about measurement uncertainty more useful and understandable if it is reported by means of confidence bands rather than as probabilistic statements about the achievement levels.

RECOMMENDATION 6.3 Achievement-level reporting should be supplemented with reporting using a standardized numeric scale. Confidence bands on this scale should be used to communicate measurement error.

As part of the reporting issues focus groups, NAGB asked parents and teachers to react to a sample score report (included as Appendix C in American Institutes for Research, 1999p). Two aspects of this sample report deserve further consideration. First, in an effort to communicate normative information, the distance between the lower and upper boundaries for each NAEP achievement level was made proportional to the percentage of students scoring at the level. The underlying scale thus resembles a percentile scale. The midpoint of this scale tends to be in the basic range. We are concerned that this format may emphasize the status quo, what students can do now, rather than the rigorous standards behind the achievement levels. Furthermore, scaling according to the distribution of students, while conveying potentially useful normative information, may lead to misinterpretations of the range of content covered by each of the achievement levels. In particular, the advanced level is represented by a very narrow box since very few students are currently at this level. One might conclude that there is relatively little material to be mastered beyond the lower boundary of the advanced range, which is clearly not the case.

A second concern with the current NAGB/AIR scheme is the continued use of the NAEP score scale for labeling both the individual point estimates and the boundaries between achievement levels. The committee is concerned that the three-digit numbers used with the NAEP scale convey a level of accuracy that is appropriate for estimates of average scores for large groups, but that is inappropriate for communicating results for individual students.

Figure 6-1 shows an example of how measurement error and resulting score uncertainty might be reported using confidence bands. In this example, we created a 41-point standard score scale with 10, 20, and 30 defining the lower boundary of the basic, proficient, and advanced levels. While 41 points may be a reasonable number of score levels, the issue of an optimal score scale cannot be decided independent of decisions about the difficulty of each test form and the accuracy of scores at different levels of achievement. It is likely, for example, that the final VNT accuracy targets will not yield the same amount of information at each achievement level.

While the focus groups conducted by AIR have provided useful information on the type of informa-

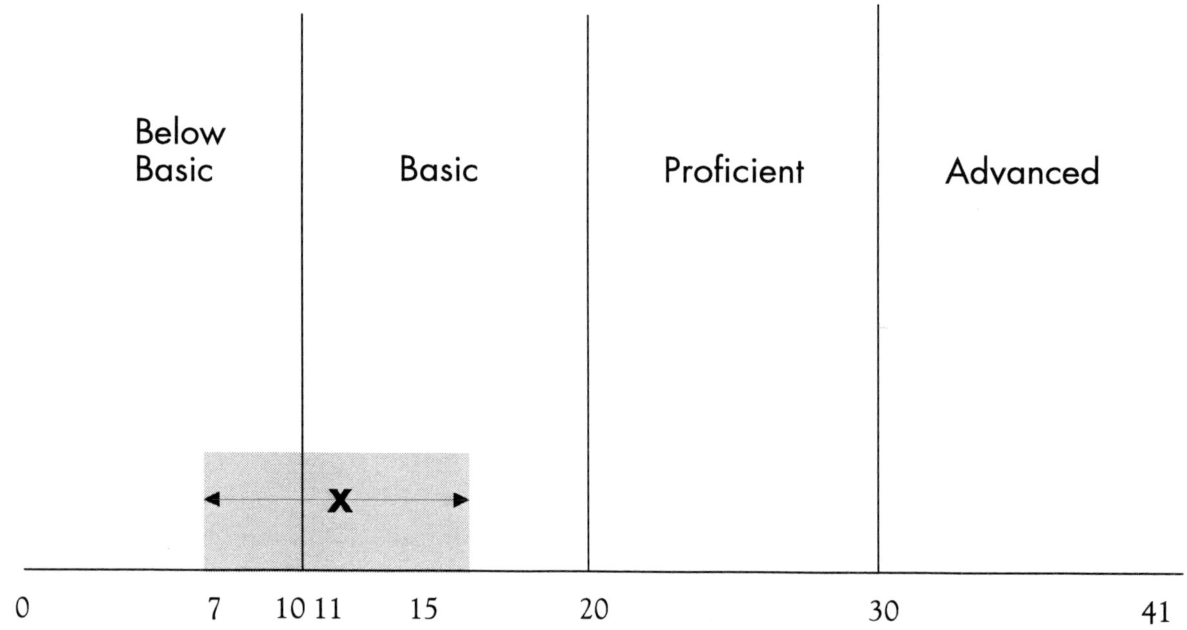

FIGURE 6-1 Sample reporting format combining achievement-level reporting with a standardized numeric scale.
NOTES: The score of 11 on this test is in the basic range. The shaded area shows the range of probable scores if the test were taken again. Statistically, scores would be expected to fall within the shaded range (between 7 and 15) 95 percent of the time.

tion that parents and teachers will understand and find useful, there is a great deal of work yet to be done. Some of the topics yet to be addressed include:

- reporting formats for the students themselves;
- the form and format in which items and student responses will be returned to students, parents, and teachers;
- what additional information about the test items (difficulty, content area, achievement level, etc.) will be provided to some or all of these audiences;
- ways in which teachers might aggregate item level results across students to identify areas of their curriculum that may need more emphasis; and
- how any test accommodations will be noted on the score reports to teachers and parents.

The committee is particularly concerned about the absence of the student's perspective on reporting issues in the work conducted to date.

Although focus groups may be useful in generating ideas, the committee is uncomfortable with basing final decisions about reporting formats only on the type of anecdotal information available from focus groups. More rigorous controlled experiments, possibly using cognitive laboratories to test understanding and use of alternative score reports, for example, might be conducted before decisions about reporting formats are reached.

The committee also realizes that preliminary decisions about reporting formats may need adjustment after the pilot test and again after the field test to reflect findings about the extent to which VNT scores can be compared with the NAEP achievement-level cutpoints. In addition, the field test should include a full operational test of reporting procedures. Results from the field test, including impact analyses, should be carefully reviewed to identify further reporting changes that might reduce sources of confusion and enhance understanding and use.

SUBSCORE REPORTING

Our call for a clearer statement of expectations for each mathematics content strand or reading stance is not meant to imply that separate scores for each strand or stance should be reported for each student. Test length considerations make it questionable whether subscores could be sufficiently reliable for individual students. A conceptual problem with subscore reporting is that the NAEP achievement levels are set for each subject as a whole. There is no established view of what it means to be proficient (or basic or advanced) within individual content areas that would define the domain of potential subscores. Unless additional achievement-level setting work is performed, subscale reporting would have to rely on an arbitrary numeric scale, supplemented by normative rather than criterion-referenced interpretive data. This concern is in addition to limitations on the reliability of subscores.

RECOMMENDATION 6.4 Individual student performance on the VNT should not be reported at the subscore level.

RECOMMENDATION 6.5 NAGB and its contractor should undertake research on alternative ways for providing item-level feedback to students, parents, and teachers. The options explored should include provision of information on item content and targeted achievement level, as well as normative information, such as passing rates.

ITEM-LEVEL INFORMATION

Perhaps the greatest distinction between the VNT and available commercial tests is the plan to release each test form immediately after operational use and to provide item-level information back to students, parents, and teachers. The developers have only begun to explore the potential value of the item-level information that will be provided. For the individual students, returning item-level information will provide a basis for diagnosing strengths and weaknesses within an overall subject domain that would be even better than subscores. For teachers, the item-level information will provide concrete illustrations of the types of problems that students should be able to solve in each content area.

There are many options for how item-level data will be provided. At one extreme, booklets and answer sheets could simply be returned as originally marked, with a separate answer key for the multiple-choice questions and scoring guides for each of the constructed-response questions. At the other extreme, items might be arranged by content area and achievement level and accompanied by normative and other psychometric data (e.g., average item scores for all students and for specific subgroups), with narrative discussions tailored for each student explaining why each incorrect option selected by the student was incorrect. There could even be summaries of the number of items in each group that the student answered correctly, perhaps with comparative normative figures. Perhaps this type of item-level information would fulfill the desire for subscores. Unlike subscores, however, this information would not consist of an arbitrary subscore scale or subscore standards, only the simple number and percentage of correct responses.

In general, the committee favors taking maximum advantage of plans to return item-level information. Information should be provided on the content area (stance or content strand) that each item was intended to measure, as well as the achievement level to which each item is linked. Since teachers and parents may try to infer what each item was intended to measure, providing this information may improve the accuracy of their inferences. Sorting the items by achievement level within content area would further aid in understanding the domain that is being measured and the expectations for student performance within this domain. Providing item difficulty information (such as the percentage of students who responded correctly) would also help parents, teachers, and students understand the student's overall score.

RECOMMENDATION 6.6 NAGB and its contractor should consider including students, particularly at the 8th-grade level, as well as parents and teachers in future focus groups on score reporting.

AGGREGATION

In its report to Congress on the purpose and uses of the VNT, NAGB adopts a stance that discourages but does not prohibit aggregation of individual student results. Specifically, (National Assessment Governing Board, 1999e:11):

> There should be no compilations of student results provided automatically by the program . . . However, it is virtually certain that compilations of student results will be desired and demanded by at least some of the state and district participants and possibly by private school participants. These participants should be permitted to obtain and compile the data at their own cost, but they will bear the full responsibility for using the data in appropriate ways and for ensuring that the uses they make of the data are valid.

The reasons for discouraging aggregation are not fully explicit. The primary concern seems to be that the aggregate results for states or large districts will not agree with NAEP results. It is possible that there is also a concern with preventing inappropriate accountability uses, since the report further states that "[NAGB] would develop and provide guidelines and criteria for use by states, districts, and schools for compiling and reporting [VNT data] in ways that are appropriate and valid" (National Assessment Governing Board, 1999e:11).

There is an apparent contradiction between the adoption of the public policy model (see Chapter 2) and this stance on aggregation. The public policy model gives the primary decision to adopt the VNT to states and districts, but they are cautioned not to aggregate the scores. Why would these entities choose to administer the VNT to their students if aggregation of information at the district and state level is discouraged?

At the committee's July meeting, it was suggested that the effort being put into preventing aggregation might be better spent explaining how the results will differ from NAEP and providing corresponding cautions in interpretation. (Yen, 1999). Some of the reasons for differences—specifically, differences in student motivation—actually suggest that aggregate VNT results might be a more valid indication of student accomplishments. Complex statistical procedures are used with NAEP to remove measurement error from estimates of differences in scores over time or among different subgroups of students. In part, these procedures are necessary because of the matrix sampling used with NAEP. Measurement error is an essential consideration because no one student takes a large and completely representative sample of items. Measurement error will be less of an issue with the VNT because each student completes a representative sample of items, which is at least twice the size of the item samples completed by students in NAEP. Indeed, the VNT is designed to provide adequate score accuracy for individual students. Most commercial tests encourage aggregation of observed student scores without requiring complex statistic machinery to remove minor biases due to measurement error.

If VNT results provide a valid assessment of the proficiency of individual students, it is perfectly legitimate for schools, districts, and even states to ask how many of their students meet these standards, or, alternatively, to track changes in means on a standard scale over time. If aggregate results are not provided to answer such questions, there may be very little motivation for districts and states to participate in the VNT.

The committee recognizes that aggregated VNT results will be likely to differ from NAEP but there could be strong advantages to explaining rather than surpressing these differences. NAGB's Linking Feasability Team (Cizek et al., 1999) identified a number of differences between the VNT and NAEP specifications. Communicating these differences to VNT users might release the VNT from excessive requirements for "NAEP-likeness." For example, NAGB believes that the calculators used with the 8th-grade mathematics test should be as much like the calculators used in NAEP as possible, even though these calculators, which in NAEP are also used in the 12th-grade, contain trigonometric functions not covered in the 8th-grade frameworks. Relaxing the "NAEP-like" requirements would mean that schools (or NAGB) could buy simple "four-function" calculators, which are much easier to explain to students unfamiliar with more advanced models and cost less than one-quarter as much. NAGB, for very good reason, has relaxed the NAEP-like requirements in the test specifications, limiting passage lengths in reading and increasing the proportion of machine-scorable items in mathematics. While these differences may limit the comparability of VNT and NAEP results, they should not be a reason to avoid supporting district- and state-level reporting of aggregated results.

RECOMMENDATION 6.7 NAGB should support aggregation of test results for participating districts and states, while discouraging inappropriate, high-stakes uses of aggregated results. NAGB should develop explicit and detailed guidelines and practices for the appropriate compilation and use of aggregate data from administration of the VNT and should explain limitations on the validity of comparisons of aggregate results on the VNT to results from NAEP.

7

Conclusions and Recommendations

In this final chapter, we recap the committee's recommendations about specific aspects of the VNT development effort discussed in detail in Chapters 2 through 6 above. These cover: test purpose and use (Chapter 2); item quality and readiness (Chapter 3); technical issues in test development (Chapter 4); inclusion and accommodation (Chapter 5); and reporting (Chapter 6).

The number of specific recommendations listed here may leave the impression that the committee is dissatisfied with the progress and pace of VNT development. As this is not necessarily the case, we end our report with two overarching conclusions about VNT development and a recommendation to Congress to consider as it decides the ultimate fate of the program.

TEST PURPOSE AND USE

The National Assessment Governing Board, in its report to Congress on VNT, has specified the purpose and intended use of the test, the meaning of "voluntary," and several other key elements:

- a focus on individual performance in reading in the 4th grade and mathematics in the 8th grade;
- an effort to link VNT content, standards, and reporting to the National Assessment of Educational Progress;
- extensive feedback of test results to individual students;
- voluntary participation by states or local or private school authorities; and
- a clearly defined prohibition of federal participation in the VNT program, beyond its support of test development and, possibly, operational costs.

The committee believes that a search for evidence that the VNT, if implemented, would have favorable effects on academic achievement should be a high priority and makes four recommendations regarding VNT purpose and use.

RECOMMENDATION 2.1 High priority should be given to the articulation of potential educational effects of the VNT and to the development of a program of research and evaluation, to determine whether and how the VNT contributes to improved educational outcomes.

RECOMMENDATION 2.2 The National Assessment Governing Board should develop explicit and detailed guidelines, practices, and enforcement mechanisms for the appropriate use of the Voluntary National Tests relative to high-stakes decisions about individual students or about teachers, classrooms, schools, or other educational units. Those guidelines should illustrate uses of the VNT relative to high-stakes decisions that are inappropriate and explicitly state the potential consequences of such inappropriate uses.

RECOMMENDATION 2.3 The National Assessment Governing Board should develop explicit and detailed guidelines, practices, and enforcement mechanisms for the appropriate compilation and use of aggregate data from administrations of the Voluntary National Tests relative to high-stakes decisions about teachers, classrooms, schools, or other educational units.

RECOMMENDATION 2.4 The National Assessment Governing Board should continue to develop plans for how the VNT would operate. Specifically, it should develop proposals for operational delivery systems for the VNT and for funding ongoing development and delivery costs so that potential users can make decisions about their participation, based on the costs as well as the potential educational value of the VNT.

ITEM QUALITY AND READINESS

The committee examined the extent to which the VNT items are likely to provide useful information to parents, teachers, students, and others about whether students have mastered the knowledge and skills specified for basic, proficient, or advanced performance in 4th-grade reading or 8th-grade mathematics.

The evaluation of the VNT items involved three key questions:

- Are the completed items judged to be as good as they can be prior to the collection and analysis of pilot test data?
- Are they likely to provide valid and reliable information for parents and teachers about students' reading or math skills?
- Does it seem likely that a sufficient number of additional items will be completed to a similar level of quality in time for inclusion in a spring 2000 pilot test?

On the basis of data from the committee's item quality rating panels and other information provided to the committee by NAGB and its contractor, the committee reached several conclusions about current VNT item quality and about the item development process:

(1) The number of items at each stage is not always known, and relatively few items and passages have been through the development and review process and fully approved for use in the pilot test.

(2) The quality of the completed items is as good as a comparison sample of released NAEP items. Item quality is significantly improved in comparison with the items reviewed in preliminary stages of development a year ago.
(3) For about half of the completed items, our experts had suggestions for minor edits, but the completed items are ready for pilot testing.
(4) Efforts by NAGB and its contractor to match VNT items to NAEP achievement-level descriptions have been helpful in ensuring a reasonable distribution of item *difficulty* for the pilot test item pool, but they have not yet begun to address the need to ensure a match of item *content* to the descriptions of performance at each achievement level.
(5) Our efforts to match item content to the achievement-level descriptions led to more concern with the achievement-level descriptions than with item content. The current descriptions do not provide a clear picture of performance expectations within each reading stance or mathematics content strand. The descriptions also imply a hierarchy among skills that does not appear reasonable to the committee.

Given these conclusions, the committee offers six recommendations.

RECOMMENDATION 3.1 NAGB should require regular item development status reports that indicate the number of items at each stage in the review process by content and format categories. For reading, NAGB should also require counts at the passage level that indicate the status of passage reviews and the completeness of all of the associated items.

RECOMMENDATION 3.2 The rates at which each of the different item types survives each stage from initial content reviews through analyses of pilot test data should be computed. This information should be used in setting targets for future item development.

RECOMMENDATION 3.3 Item quality concerns identified by reviewers, such as distractor quality and other "minor edits," should be carefully addressed and resolved by NAGB and its contractor prior to inclusion of any items in pilot testing.

RECOMMENDATION 3.4 The contractor should continue to refine the achievement-level matching process to include the alignment of item *content* to achievement-level descriptions, as well as the alignment of item *difficulty* to the achievement-level cutpoints.

RECOMMENDATION 3.5 The achievement-level descriptions should be reviewed for usefulness in describing specific knowledge and skill expectations to teachers, parents, and others with responsibility for interpreting test scores and promoting student achievement.

RECOMMENDATION 3.6 Test blueprints should be expanded to indicate the expected number of items at each achievement level for each content area (reading stance or mathematics content strand) for each form of the test. Insofar as possible, items at each achievement level should be included for each content area.

TECHNICAL ISSUES IN TEST DEVELOPMENT

Our year 2 evaluation of technical issues in test development focused on the extent to which the design for pilot testing will result in items that represent the content and achievement-level specifications, are free of bias, and support test form assembly; plans for the implementation of VNT pilot testing; plans for assembling field test forms likely to yield valid achievement-level results; and the technical adequacy of revised designs for field testing, equating, and linking.

NAGB and its contractor have made progress in developing detailed plans for score reporting, the design and analysis of pilot test data to screen items for inclusion in VNT forms, and on the difficult issues associated with the field test. Based on information available to date, we offer six recommendations.

RECOMMENDATION 4.1 Pilot test plans should include school clusters, overlapping (hybrid) forms design, and NAEP anchor forms, as currently planned. In addition, the contractor should select the calibration procedure that is best suited to the final data collection design and in accord with software limitations and should plan to conduct item-fit analyses.

RECOMMENDATION 4.2 Information regarding expected item survival rates from pilot to field test should be stated explicitly, and NAGB should consider pilot testing additional constructed-response items, given the likelihood of greater rates of problems with these types of items than with multiple-choice items.

RECOMMENDATION 4.3 NAGB and its contractor should continue to detail plans for analyzing the pilot test data. Additional specifications should be provided for assessing the extent to which each item fits the model being used for calibration and the ways in which differential item functioning analyses results will be used in making decisions about the items.

RECOMMENDATION 4.4 A target test information function should be decided on and set. Although accuracy at all levels is important, accuracy at the lower boundaries of the basic and proficient levels appears most critical. Equivalent accuracy at the lower boundary of the advanced level may not be feasible with the current mix of items, and it may not be desirable because the resulting test would be too difficult for most students.

RECOMMENDATION 4.5 NAGB should consider plans for developing an alternate form of the VNT targeted to students at the low end of the achievement scale.

RECOMMENDATION 4.6 Plans for the VNT pilot test should include efforts to gather empirical data on the effects of content, administration, and use differences between the VNT and NAEP on the feasibility of linking VNT scores to the NAEP score scale. Specifically, a NAEP-like form (e.g., two non-overlapping booklets from recent 4th-grade reading and 8th-grade mathematics assessments) should be included to allow for an assessment of the effect of content differences and administration differences on the linkage of VNT scores to the NAEP scale.

INCLUSION AND ACCOMMODATION

There are two key challenges to testing students with disabilities or limited English proficiency. The first is to establish effective procedures for identifying and screening such students so they can appropriately be included in assessment programs. The second is to identify and provide necessary accommodations to students with special needs while maintaining comparable test validity with that for the general population.

The committee applauds AIR's proposal to evaluate the effects of two common accommodations on VNT performance (i.e., extended time and small-group administrations), among students with disabilities and with limited English proficiency in the pilot test. In addition, proposed research on extra time and small-group or one-on-one administration in conjunction with the pilot test have been approved. However, no specific recommendations or actions appear to have been made or taken on the basis of the hearings on inclusion and accommodation, and parent and teacher focus groups did not specifically address these issues.

While language simplification methodology has been used in the test development process, little attention has been paid to other language issues, aside from dual-language booklets, regarding the VNT mathematics test (e.g., whether translation of existing questions into Spanish and other language versions will produce comparable items or whether methods can be used to reduce the reading level of mathematics items). Participation in the cognitive laboratories by students with disabilities and with limited English proficiency has been expanded. On the basis of the work done so far, the committee offers five recommendations.

RECOMMENDATION 5.1 NAGB should accelerate its plans, research, and schedule for inclusion and accommodation of students with disabilities and limited English proficiency in order to increase the participation of both those student populations in numbers representative of their numbers in the student population.

RECOMMENDATION 5.2 NAGB should consider expanding the accommodation research planned in conjunction with the pilot test to include a systematic analysis of the use and effect of dual-language booklets. Additional accommodations for English-language learners, in the forms of both a Spanish-only translation of the mathematics test and the use of English-Spanish and English-other languages dictionaries for the mathematics test, should also be considered for the pilot test.

RECOMMENDATION 5.3 NAGB should clarify the reading constructs (e.g., reading proficiency, reading proficiency in English, etc.) being measured by the 4th-grade reading test prior to the field test and then address what accommodations would not invalidate assessment of these constructs. In particular, NAGB should clarify when reading competency could be assessed in a student's primary or native language if it is not English.

RECOMMENDATION 5.4 The National Assessment Governing Board should assess the effects of various accommodations for limited-English-proficient students and students with disabilities at both the item and total test score levels. To do so will require oversampling in the pilot and field tests.

RECOMMENDATION 5.5 The National Assessment Governing Board should provide a clear, concise, and detailed list of accommodations for the VNT for students with disabilities or limited English proficiency for use on the VNT field test.

REPORTING

One of the primary recommendations of the NRC year 1 report was that decisions about how scores will be computed and reported should be made before the design of the VNT test forms can be fully evaluated. NAGB and AIR are developing and evaluating options for test use and have conducted focus groups that have included consideration of reporting options, but no further decisions about score computation and reporting have been made. The committee believes that a number of decisions and steps related to VNT reporting need to be made soon.

RECOMMENDATION 6.1 Given that test items and answer sheets will be provided to students, parents, and teachers, as well as made available to the general public, test forms should be designed to support scoring using a straightforward, total correct, raw score approach.

RECOMMENDATION 6.2 Special attention should be given to the work required for receiving partial credit for constructed-response items that have full scores of more than 1 point.

RECOMMENDATION 6.3 Achievement-level reporting should be supplemented with reporting using a standardized numeric scale. Confidence bands on this scale should be used to communicate measurement error.

RECOMMENDATION 6.4 Individual student performance on the VNT should not be reported at the subscore level.

RECOMMENDATION 6.5 NAGB and its contractor should undertake research on alternative ways for providing item-level feedback to students, parents, and teachers. The options explored should include provision of information on item content and targeted achievement level, as well as normative information, such as passing rates.

RECOMMENDATION 6.6 NAGB and its contractor should consider including students, particularly at the 8th-grade level, as well as parents and teachers in future focus groups on score reporting.

RECOMMENDATION 6.7 NAGB should support aggregation of test results for participating districts and states, while discouraging inappropriate, high-stakes uses of aggregated results. NAGB should develop explicit and detailed guidelines and practices for the appropriate compilation and use of aggregate data from administration of the VNT and should explain limitations on the validity of comparisons of aggregate results on the VNT to results from NAEP.

SUMMARY CONCLUSIONS AND RECOMMENDATION

Congress must answer the overarching policy question about the VNT: whether development should continue or be terminated. The committee does not take a position either for or against continued development. Our primary charge was to evaluate the technical quality of the VNT test items and forms, and our recommendations above offer a number of specific ways to improve the VNT items and the development process more generally.

Lest these recommendations be misinterpreted as a condemnation of the quality of VNT development, however, we stress that there is no evidence that the current process should be halted on technical grounds and we offer the following summary conclusion:

CONCLUSION VNT development is generally on course. A large number of items have been written, and the quality of the items that came through the contractor's development and review process is comparable to the quality of items from NAEP. Plans for pilot testing these items are generally sound.

Rather than either approving or terminating the VNT, Congress could elect to postpone a final decision until a more considered debate of value and costs is completed. If an overall decision is deferred, Congress must also decide whether to allow or continue to prohibit the collection of pilot test data. In this context, we offer our second general conclusion:

CONCLUSION The planned pilot test of VNT test items presents opportunities for research on a number of important test development topics that will be useful to NAEP and state and local assessment programs even if the VNT is eventually terminated. These research opportunities include: (1) assessing the quality and effectiveness of the VNT's item development, review, and revisions processes; (2) collecting empirical data on the impact of different threats to "linkability"; and (3) assessing the feasibility, effects, and validity of alternative testing accommodations for students with disabilities or limited English proficiency. In addition, the items themselves are likely to be useful for other testing programs.

This second conclusion is not meant to imply that the pilot test is a good idea if a decision has already been reached to terminate the program. Rather, we suggest that, if a decision to terminate the program is deferred pending further consideration of its potential value, there could be value to going ahead with the pilot test in parallel with reaching a final conclusion about the VNT itself. As stated by NAGB in its report on the VNT's purpose and use report, scores from the pilot test would not be reported back to students, parents, or teachers so the risk to students is minimal.

The key question would be whether the costs, including the time of teachers and students, justify the potential benefits. The intended value of the VNT is to improve the achievement of American students at key points in their educational careers. However, the consequential chain from reporting the achievement of individual students relative to challenging national standards through parent involvement, student behavior, and teacher practices to improved achievement has not been carefully delineated. And questions of who will pay ongoing development and administration costs have not been answered.

We began this report with a discussion of the proposed purpose and use of the VNT. NAGB has articulated a viable purpose for the VNT, but Congress alone can assess the "value" that would result if the VNT is implemented as described in NAGB's purpose and use document. The committee does not recommend either continuation or termination, but we do offer a recommendation to Congress in making the decision:

RECOMMENDATION TO CONGRESS The decision to continue or terminate the VNT should be based on a carefully articulated statement of the expected value and costs of the program, including a detailed examination of underlying assumptions and a delineation of possible unintended outcomes. To the maximum extent possible, research on results from other educational reform efforts should be considered to support or contradict assumptions in this value-and-cost statement. Information on the likelihood of use by states, districts, and individuals should also be considered in making a decision about the VNT.

References

American Educational Research Association, American Psychological Association, and National Council on Measurement in Education

in press *Standards for Educational and Psychological Testing*. Washington, DC: American Educational Research Association.

American Institutes for Research

1998a Background Paper Reviewing Laws and Regulations, Current Practice, and Research Relevant to Inclusion and Accommodations for Students with Disabilities. Prepared for the National Assessment Governing Board in Support of Contract RJ97153001. Washington, DC, July 23.

1998b Background Paper Reviewing Laws and Regulations, Current Practice, and Research Relevant to Inclusion and Accommodations for Students with Disabilities. Prepared for the National Assessment Governing Board in Support of Contract RJ97153001. Palo Alto, CA, November 6.

1998c Background Paper Reviewing Laws and Regulations, Current Practice, and Research Relevant to Inclusion and Accommodations for Students with Limited English Proficiency. Prepared for the National Assessment Governing Board in Support of Contract RJ97153001. Palo Alto, CA, November 6.

1998d Cognitive Lab Report. Prepared for the National Assessment Governing Board in Support of Contract RJ97153001. Washington, DC, July 29.

1998e Designs and Item Calibration Plans for Including NAEP Item Blocks in the 1999 Pilot Test of the VNT. Washington, DC, September 1.

1998f Designs and Item Calibration Plan for the 1999 Pilot Test. Washington, DC, July 24.

1998g Linking the Voluntary National Tests with NAEP and TIMSS: Design and Analysis Plans. Washington, DC, February 20.

1998h List of Groups and Agencies with Interests in Issues Related to Inclusion and Accommodations for Students with Disabilities. Prepared for the National Assessment Governing Board in Support of Contract RJ97153001. Palo Alto, CA, November 6.

1998i List of Groups and Agencies with Interests in Issues Related to Inclusion and Accommodations for Students with Limited English Proficiency. Prepared for the National Assessment Governing Board in Support of Contract RJ97153001. Palo Alto, CA, November 6.

1998j Proposed Plan for Calculator Use. Washington, DC, July 23.

1998k Revised Inclusion and Accommodations Workplan. Prepared for the National Assessment Governing Board in Support of Contract RJ97153001. Washington, DC, May 8.

1998l Score Reporting, Draft. Washington, DC, October 20.

1998m Summary of VNT Activities and Plans Related to the Assessment of Students with Disabilities and Students with Limited English Proficiency. Prepared for the National Assessment Governing Board in Support of Contract RJ97153001. Palo Alto, CA, November 6.

1998n Test Utilization, Draft. Washington, DC, October 13.

1998o VNT: Report on Scoring Rubric Development.

1999a Cognitive Lab Report: Lessons Learned. Prepared for the National Assessment Governing Board in Support of Contract RJ97153001. Washington, DC, March 23.

1999b Content Coverage and Achievement Level Reviews of Items Developed in 1997-1998. Prepared for the National Assessment Governing Board in Support of Contract RJ97153001. Washington, DC, February 3.

1999c Field Test Plans for the VNT: Design and Equating Issues. Scott Oppler and Steve Ferrara. Unpublished document presented at February 25 meeting of Committee on the Evaluation of the VNT, Year 2.

1999d Flowchart of VNT New Item Production Process. Unpublished document presented at Feb. 25 meeting of Committee on the Evaluation of the VNT, Year 2. February 11.

1999e Proposed Year 3-5 Research Plan. Prepared for the National Assessment Governing Board in Support of Contract RJ97153001. Washington, DC, April 28.

1999f Report on the Status of the Voluntary National Tests Item Pools as of March 1, 1999. Prepared for the National Assessment Governing Board in Support of Contract RJ97153001. Washington, DC, March 30.

1999g Score Reporting, Scoring Examinees, and Technical Specifications: How Should These Be Influenced by the Purposes and Intended Uses of the VNT? March 31.

1999h Selected Item Response Theory Scoring Options for Estimating Trait Values. Wendy Yen. Unpublished document presented at AIR TAC meeting. April 8.

1999i Technical Specifications, Revisions as of June 18, 1999.

1999j Training Materials for VNT Protocol Writing, February/March.

1999k VNT: Counts of Reading Passages Using Revised Taxonomies, June 24, 1999.

1999l VNT: Forms Assembly Procedures and Technical Specifications, July 20, 1999.

1999m VNT in Mathematics; Proposed Outline for the Expanded Version of the Test Specifications, February 2, 1999.

1999n VNT in Reading: Proposed Outline for the Expanded Version of the Test Specifications, February 12, 1999.

1999o VNT Interviewer Training Manual.

1999p VNT: Issues Concerning Score Reporting for the Voluntary National Tests: Results of Parent and Teacher Focus Groups, April 27.

1999q VNT: Plans for Continuing Work in Score Reporting, July 27.

1999r VNT: Proposed Score Reporting Metrics and Examinee Scoring Algorithms for the Voluntary National Tests, July 20.

Anderson, R.C., and P.D. Pearson

1984 A schema-theoretic view of basic processes in reading. Pp. 255-291 in *Handbook of Reading Research*, P.D. Pearson, R. Barr, M.L. Kamil, and P. Mosenthal, eds. New York: Longman.

Cizek, G.J., P.A. Kenney, M.J. Kolen, C.W. Peters, and W.J. Van der Linden

1999 Final Report of the Study Group Investigating the Feasibility of Linking Scores on the Proposed Voluntary National Tests and the National Assessment of Educational Progress, May draft. Study commissioned by the National Assessment Governing Board. University of Toledo.

Donahue, P.L., K.E. Voelkl, J.R. Campbell, and J. Mazzeo

1999 *The NAEP 1998 Reading Report Card for the Nation*. NCES 1999-459. Washington DC: U.S. Department of Education, Office of Educational Research and Improvement, National Center for Education Statistics.

Ercikan, Kadriye

1999 Synthesis Paper on VNT Pilot Test Design Features, June 16. Paper commissioned by American Institutes for Research. University of British Columbia, Canada.

Graesser, A., M. Singer, and T. Trabasso

1994 Constructing inferences during narrative text comprehension. *Psychological Review* 101:371-395.

Guerra, Michael

1998 The National Assessment Governing Board and Voluntary National Tests: A Status Report. November 19, 1998. National Assessment Governing Board, Washington, DC.

Hambleton, R.K., R.L. Brennan, W. Brown, B. Dodd, R.A. Forsyth, W.A. Mehrens, J. Nelhaus, M. Reckase, D. Rindone, W.J. van der Linden, R. Zwick

1999 A Response to "Setting Reasonable and Useful Performance Standards" in *Grading the Nation's Report Card*. Paper presented to the National Assessment Governing Board, May 13. University of Massachusetts, Amherst.

Hanson, Brad
1999 Evaluation of VNT Pilot Test Design, May 7. Paper commissioned by American Institutes for Research. ACT, Inc., Iowa City, IA.

Hoffman, R.G., and A.A. Thacker
1999 Evaluation of the Quality of the Voluntary National Test Item Pool: Content, Craftsmanship, and Achievement Level. Prepared for the Committee on the Evaluation of the Voluntary National Tests, Year 2 in Support of Contract EDUC-2487-99-001. May. Human Resources Research Organization, Alexandria, VA.

Johnson, Eugene G.
1999a Revised Plans for Linking the Voluntary National Test with NAEP, June 11. Educational Testing Service, Princeton, NJ.
1999b VNT Pilot Design Features, May 28. Educational Testing Service, Princeton, NJ.

Kintsch, W.
1988 *Comprehension: A Paradigm for Cognition.* Cambridge, England: Cambridge University Press.

Lorch, R.F., and R. van den Broek
1997 Understanding reading comprehension: Current and future contributions of cognitive science. *Contemporary Educational Psychology* 22(4):213-247.

Mazzeo, J.
1999 Increasing Participation of Special Needs Students in NAEP: Results of the 1996 Research Study. Unpublished document presented at the National Assessment Governing Board meeting. August 7. Educational Testing Service, Princeton, NJ.

National Academy of Education
1993 Setting Performance Standards for Student Achievement, Robert Glaser, Robert Linn, and George Bohrnstedt, eds. Panel on the Evaluation of the NAEP Trial State Assessment. Stanford, CA: National Academy of Education.

National Assessment Governing Board
1998a Voluntary National Test: Inclusions and Accommodations for Test Development. Policy Statement. Draft, July 7.
1998b Voluntary National Test in 4th Grade Reading: Test Specifications Outline. March 7.
1998c Voluntary National Test in 8th Grade Mathematics: Test Specifications Outline. March 7.
1999a Overview: Determining the Purpose, Intended Use, Definition of the Term Voluntary, and Reporting for the Proposed Voluntary National Test.
1999b Public Hearings and Written Testimony on Students with Disabilities and the Proposed Voluntary National Test: October-November 1998. Synthesis Report. Draft, February.
1999c Public Hearings and Written Testimony on Students with Limited English Proficiency and the Proposed Voluntary National Test: October-November 1998. Synthesis Report. Draft, February.
1999d Public Hearings and Written Testimony on the Purpose, Intended Use, Definition of the Term "Voluntary," and Reporting of the Proposed Voluntary National Test. Synthesis Report. April.
1999e The Voluntary National Test: Purpose, Intended Use, Definition of Voluntary and Reporting. Adopted unanimously by the National Assessment Governing Board, June 23.

National Research Council
1997 *Educating One and All: Students with Disabilities and Standards-Based Reform.* Lorraine M. McDonnell, Margaret J. McLaughlin, and Patricia Morison, eds. Committee on Goals 2000 and the Inclusion of Students with Disabilities, Board on Testing and Assessment, Commission on Behavioral and Social Sciences and Education. Washington, DC: National Academy Press.
1998a Letter to Secretary Richard Riley, U.S. Department of Education and Mark D. Musick, National Assessment Governing Board, from Robert M. Hauser and Lauress L. Wise, co-principal investigators, Evaluation of the Voluntary National Test. Board on Testing and Assessment, Commission on Behavioral and Social Sciences and Education. Washington, DC.
1998b *Preventing Reading Difficulties in Young Children.* Catherine E. Snow, M. Susan Burns, and Peg Griffin, eds. Committee on the Prevention of Reading Difficulties in Young Children, National Research Council. Washington, DC: National Academy Press.
1999a *Grading the Nation's Report Card: Evaluating NAEP and Transforming the Assessment of Educational Progress.* Committee on the Evaluation of National and State Assessments of Educational Progress, James W. Pellegrino, Lee R. Jones, and Karen J. Mitchell, eds. Board on Testing and Assessment, Commission on Behavioral and Social Sciences and Education. Washington, DC: National Academy Press.

1999b *Evaluation of the Voluntary National Tests, Phase I Report.* Lauress L. Wise, Robert M. Hauser, Karen J. Mitchell, and Michael J. Feuer, eds. Project on Evaluation of the Voluntary National Test. Board on Testing and Assessment, Commission on Behavioral and Social Sciences and Education. Washington, DC: National Academy Press.

1999c *Evaluation of the Voluntary National Tests, Year 2: Interim Report.* Committee on the Evaluation of the Voluntary National Tests, Year 2, Lauress L. Wise, Richard J. Noeth, and Judith A. Koenig, eds, Board on Testing and Assessment, Commission on Behavioral and Social Sciences and Education. Washington, DC: National Academy Press.

1999d *High Stakes: Testing for Tracking, Promotion, and Graduation.* Committee on Appropriate Test Use, Jay P. Heubert and Robert M. Hauser, eds. Board on Testing and Assessment, Commission on Behavioral and Social Sciences and Education. Washington, DC: National Academy Press.

1999e *Uncommon Measures: Equivalence and Linkage Among Educational Tests.* Committee on Equivalency and Linkage of Educational Tests, Michael J. Feuer, Paul W. Holland, Bert F. Green, Meryl W. Bertenthal, and F. Cadelle Hemphill, eds. Board on Testing and Assessment, Commission on Behavioral and Social Sciences and Education, Washington DC: National Academy Press.

National Research Council and Institute of Medicine

1997 *Improving Schooling for Language-Minority Children: A Research Agenda.* Diane August and Kenji Hakuta, eds. Committee on Developing a Research Agenda on the Education of Limited-English-Proficient and Bilingual Students, Board on Children, Youth, and Families, National Research Council and Institute of Medicine. Washington, DC: National Academy Press.

Paulsen, Christine

1999 Using Social Moderation to Link the VNT to NAEP, June 18. American Institutes for Research, Washington, DC.

Popham, James W.

1990 *Modern Educational Measurement: A Practitioner's Perspective.* Needham, MA: Allyn and Bacon.

Reckase, Mark

1999 An Evaluation of the VNT Pilot Test Design, May 12. Paper commissioned by American Institutes for Research. Michigan State University.

Riley, Richard

1997 Letter to the editor. *Washington Post* August 30:A26.

Shaughnessy, C.A., J.E. Nelson, and N.A. Norris

1997 *NAEP 1996 Mathematics Cross-State Data Compendium for the Grade 4 and Grade 8 Assessment.* Washington DC: National Center for Education Statistics.

Yen, Wendy

1999 Brief Comments on the Linking Feasibility Team Report, July 31. CTB/McGraw Hill, Monterey, CA.

APPENDIX A

The National Assessment Governing Board's Draft Scenarios for the Purpose and Use of the Voluntary National Tests

Draft Scenarios for the Proposed Voluntary National Test

<table>
<tr><th></th><th>Public Policy Model</th><th>Individual Decision Model</th></tr>
<tr><td>Purpose</td><td colspan="2">To measure individual student achievement in 4th grade reading and 8th grade mathematics, based on the rigorous content and rigorous performance standards of the National Assessment of Educational Progress (NAEP), as set by the National Assessment Governing Board (NAGB).</td></tr>
<tr><td>Voluntary
(Federal Role)</td><td colspan="2">The federal government shall not require participation by any state, district, public or private school, organization or individual in voluntary national tests or require participants to report voluntary national test results to the federal government.</td></tr>
<tr><td>Voluntary
(Who decides)</td><td>• Public and private school authorities volunteer
• State and/or local law and policy determines decision level (i.e., public policy model begins at the state level, then proceeds through district, and school--see Overview for description)
• Parents "opt out" as determined by state/local law and policy</td><td>• Parents decide whether student participates</td></tr>
<tr><td>Intended Use</td><td>To provide information to parents, students, and authorized educators about the achievement of the individual student in relation to rigorous content and rigorous performance standards based on NAEP, as set by NAGB.</td><td>To provide information to parents and students about the child's achievement in relation to rigorous content and rigorous performance standards based on NAEP, as set by NAGB.</td></tr>
<tr><td rowspan="2">Reporting</td><td colspan="2">• Results reported by NAEP performance standards (i.e., achievement levels--Basic, Proficient, Advanced)
• Explanation of achievement levels in light of test questions taken by student
• All test questions, student answers, and answer key returned in timely fashion
• Easy to understand, readable</td></tr>
<tr><td>• Parents, students, and authorized educators receive reports
• Some norm-referenced information (e.g., percent of students nationally at each achievement level, taken from the field test results)
• No aggregate data will be provided automatically (i.e., by class, school, district, and state), but individual data can be compiled by state/local participants, who will bear responsibility for using resulting data in valid, appropriate ways
• Guidance provided on technical criteria for aggregate reporting if done by participants</td><td>• Parents and students receive reports
• Some norm-referenced information (e.g., percent of students nationally at each achievement level taken from the field test results), but no comparisons at class, school, district, or state levels</td></tr>
</table>

Appendix: Implementation and Other Issues

<table>
<tr><th></th><th>Public Policy Model</th><th>Individual Decision Model</th></tr>
<tr><td rowspan="2">Possible uses by others*</td><td colspan="2">• General indicator of individual achievement against rigorous external standards established through a national consensus process</td></tr>
<tr><td>• Parent/teacher follow up recommended but decided at state/district/school as appropriate
• Results can be compared to student performance on state and/or local tests as a basis for examining the content of state/local standards
• Local decision to use as one of several criteria about individual student; should be validated
• States may want to use as an external anchor to their state tests
• Since only one grade/two subjects, not much information for use as part of school accountability system; any such use should be validated</td><td>• Follow up with school/ teacher is up to the parent</td></tr>
<tr><td>The VNT is Not</td><td colspan="2">• It is NOT tied to a preferred curriculum, teaching method or approach
• It is NOT intended for diagnosing specific learning problems or English language proficiency
• It is NOT intended as sole criterion in high stakes decision about individual student
• It is NOT intended for evaluating instructional practices, programs, or school effectiveness</td></tr>
<tr><td rowspan="2">Possible Test Delivery Models</td><td colspan="2"><u>Central Management and Oversight</u>: A federal agency takes the VNT as developed by the Governing Board; develops policies for quality control, security and and reporting; contracts for printing, testing, scoring and reporting services; disseminates information about the test schedule; handles the "sign-up" of participants; monitors the testing; and ensures the quality control of results.</td></tr>
<tr><td colspan="2"><u>Free Market Model</u>: The VNT is developed by NAGB, licensed for marketing by commercial test publishers, and marketed like any commercial test for use by any appropriate public or private educational agency, testing center, or individual. Parents may "opt out" as determined by state law and policy and may "opt in" by purchasing private testing services if the test is not offered at their child's school. Quality control monitoring, rigor of test security, training of test administrators, content of reports, development of "non-standard" versions of tests, use of norms, etc., determined by costs and market.</td></tr>
<tr><td>Administration</td><td>• Dissemination strategy to public and private education decision makers
• Testing in participating schools
• Training of test administrators
• Testing during specified date in March
• Quality control monitoring of testing
• Guidance to teachers on appropriate test preparation practices
• Reports sent to states, districts, schools, teachers and parents per state/local policy</td><td>• Similar to SAT/ACT "Self-select" model
• Dissemination strategy to parents
• Parents sign-up at cooperating schools/test centers
• Testing at cooperating schools/test centers
• Testing during specified date in March
• Quality control monitoring of testing
• Reports sent to parents
• Q&A system available for parents</td></tr>
<tr><td rowspan="3">Who Pays: Three Options</td><td colspan="2"><u>Option 1:</u>Federal Gov't pays all costs: test development, testing, scoring & reporting.</td></tr>
<tr><td colspan="2"><u>Option 2:</u>Fed. Gov't pays for test development; volunteer (whether state, district, school, or parent) pays for testing, scoring & reporting.</td></tr>
<tr><td colspan="2"><u>Option 3:</u>Fed. Gov't pays all costs initially; volunteer pays for all costs but development after year 1</td></tr>
<tr><td rowspan="2">Possible Consequences
<u>Positive:</u>

<u>Negative:</u></td><td colspan="2">• Parents become more involved with child's education
• Students study harder and learn more
• Teachers work more to emphasize important skills and knowledge in the subjects tested
• Parents, students, and teachers have a means for better communications about the child's achievement</td></tr>
<tr><td colspan="2">• VNT test-preparation "industry" for economically advantaged students
• Inappropriate test preparation practices and over-emphasis on test-taking techniques
• Misuse of test results
• Cheating scandals; security breaches
• Litigation against NAGB</td></tr>
</table>

* This list is intended to be illustrative, not exhaustive, of uses that can be imagined that others may want to make of the VNT. Any use of the VNT beyond the intended use described in the draft scenarios should be validated for its applicability and appropriateness by the respective user.

APPENDIX

B

Achievement-Level Descriptions For 4th-Grade Reading and 8th-Grade Mathematics

NAEP 4th-GRADE READING ACHIEVEMENT LEVEL DESCRIPTIONS

Level	*General*	*Literary Texts*	*Informational Texts*
Basic	• Demonstrate an understanding of the overall meaning of what they read • Make relatively obvious connections between the text and their own experience	• Tell what the story is generally about • Provide details to support their understanding • Connect aspects of the story to their own experience	• Tell what the text is generally about or identify the purpose for reading it • Provide details to support their understanding • Connect ideas from the text to their own background knowledge and experiences
Proficient	• Demonstrate an overall understanding of the text, providing inferential as well as literal information • Extend their ideas by making inferences, drawing conclusions, and making connections to their own experience • The connection between the text and what the student infers should be clear	• Summarize the story • Draw conclusions about characters or plot • Recognize relationships such as cause and effect	• Summarize information and identify authors intent or purpose • Draw reasonable conclusions from the text • Recognize relationships such as cause and effect or similarities and differences • Identify the meaning of the selection's key concepts
Advanced	• Generalize about topics in the reading selection • Demonstrate an awareness of how authors compose and use literary devices • Judge texts critically • Give answers that indicate careful thought	• Make generalizations about the point of the story and extend its meaning by integrating personal experiences and other readings with the ideas suggested by the text • Identify literary devices such as figurative language	• Explain the author's intent by using supporting material from the text • Make critical judgments of the form and content of the text • Explain their judgments clearly

NAEP 8th-GRADE MATHEMATICS ACHIEVEMENT LEVEL DESCRIPTIONS

Level	*General*	*Specifics*
Basic	• Conceptual and procedural understanding of the five NAEP content strands • Understanding of arithmetic operations—including estimation—on whole numbers, decimals, fractions, and percents	• Complete problems correctly with the help of structural prompts such as diagrams, charts and graphs • Solve problems in all NAEP content strands through the appropriate selection and use of strategies and technological tools—including calculators, computers, and geometric shapes • Use fundamental algebraic and geometric concepts in problem solving • As they approach the proficient level, students should be able to determine which of the available data are necessary and sufficient for correct solutions and use them in problem solving. • However, these 8th graders show limited skill in communicating mathematically
Proficient	• Apply mathematical concepts and procedures consistently to complex problems in the five NAEP content strands	• Conjecture, defend their ideas, and give supporting examples • Understand the connections between fractions, percents, decimals, and other mathematical topics such as algebra and functions • Have a thorough understanding of basic-level arithmetic operations—an understanding sufficient for problem solving in practical situations • Quantity and spatial relationships in problem solving and reasoning should be familiar • Convey underlying reasoning skills beyond the level of arithmetic • Compare and contrast mathematical ideas and generate their own examples. • Make inferences from data and graphs • Apply properties of informal geometry • Accurately use the tools of technology • Understand the process of gathering and organizing data • Be able to calculate, evaluate, and communicate results within the domain of statistics and probability
Advanced	• Reach beyond the recognition, identification, and application of mathematical rules in order to generalize and synthesize concepts and principals in the five NAEP content strands	• Probe examples and counterexamples in order to shape generalizations from which they can generate models • Use number sense and geometric awareness to consider the reasonableness of an answer • Use abstract thinking to create unique problem-solving techniques and explain the reasoning processes underlying their conclusions

SOURCE: Hoffman and Thacker (1999:14-15).